# GLOBES

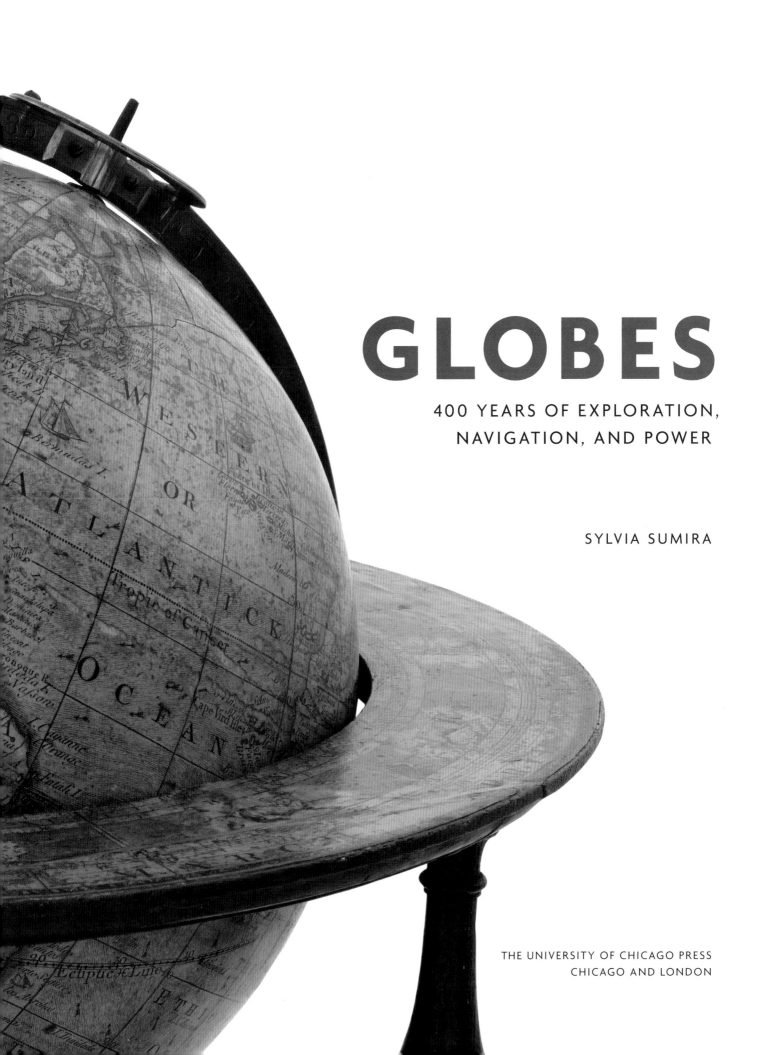

# GLOBES

## 400 YEARS OF EXPLORATION, NAVIGATION, AND POWER

SYLVIA SUMIRA

THE UNIVERSITY OF CHICAGO PRESS
CHICAGO AND LONDON

To Jim

Sylvia Sumira is a leading authority on historic globes and one of few
conservators in the world to specialize in printed globes. She worked at
the National Maritime Museum, Greenwich, before setting up her own
studio, where she carries out conservation work for museums, libraries,
and other institutions, as well as for private owners.

The University of Chicago Press, Chicago 60637
The University of Chicago Press, Ltd., London
Text © 2014 by Sylvia Sumira
Illustrations © 2014 by The British Library Board
    and other named copyright holders
All rights reserved. Published 2014.
Printed in Hong Kong

23 22 21 20 19 18 17 16 15 14     1 2 3 4 5

ISBN-13: 978-0-226-13900-5 (cloth)
ISBN-13: 978-0-226-13914-2 (e-book)
DOI: 10.7208/chicago/9780226139142.001.0001

Library of Congress Cataloging-in-Publication Data

Sumira, Sylvia, author.
    Globes : 400 years of exploration, navigation, and power / Sylvia Sumira.
        pages cm
    Includes bibliographical references.
    ISBN 978-0-226-13900-5 (cloth : alk. paper) — ISBN 978-0-226-13914-2
    (e-book) 1. Globes—History. 2. Cartography—History. I. Title.
    G3170.S955 2014
    912—dc23
                    2013031279

♾ This paper meets the requirements of ANSI/NISO Z39.48-1992
    (Permanence of Paper).

# CONTENTS

# PREFACE

This book introduces the subject of globes from the start of the sixteenth to the end of the nineteenth century, the period when they were most widely used. It concentrates on printed terrestrial and celestial globes, but a few other types are included in order to show particularly significant or remarkable examples. The book takes a chronological approach, and draws largely on the British Library's impressive collection, filling notable gaps with globes from other sources. Not all makers of globes are included here: for those inspired to find out more, the bibliography is a useful starting point.

A note on the dimensions of globes may be helpful. Globes were made with specific diameters, but the units of measurement have not always been standard throughout Europe. In the British Isles globe-makers gave the diameter of their globes in inches, but this is not always the case for globes made outside Britain, For completeness, measurements are provided in both inches and centimetres throughout this volume.

The glossary explains technical terms with which the reader may be unfamiliar.

There are several people whom I wish to acknowledge for their support during the writing of this book. At the British Library I would like to thank Peter Barber, Head of the Map Library, for his advice and encouragement, in addition to his very generous help over many years, and Tom Harper, Curator of Antiquarian Maps, for his invaluable assistance when looking at globes. Thanks must also be given to Geoff Armitage and Nicola Beech for their friendly and knowledgeable help while I was navigating the Map Room, to Elizabeth Hunter for photography, and to Lara Speicher, Sally Nicholls and Robert Davies from British Library Publishing, and to Maggi Smith, the designer. Though many of her works are listed in the bibliography, I feel special thanks must be given to Elly Dekker for her generosity. Her studies of globes have been a source of knowledge and inspiration for many years. Thanks also to Kristen Lippincott for her thoughtful comments and suggestions, to Paul Cook at the National Maritime Museum for sharing his insights about globes with me, and to Rosemarie Sumira for her support in a myriad of ways. Finally, massive thanks must go to Jim Bennett, my husband, for his patience, encouragement and constructive criticism.

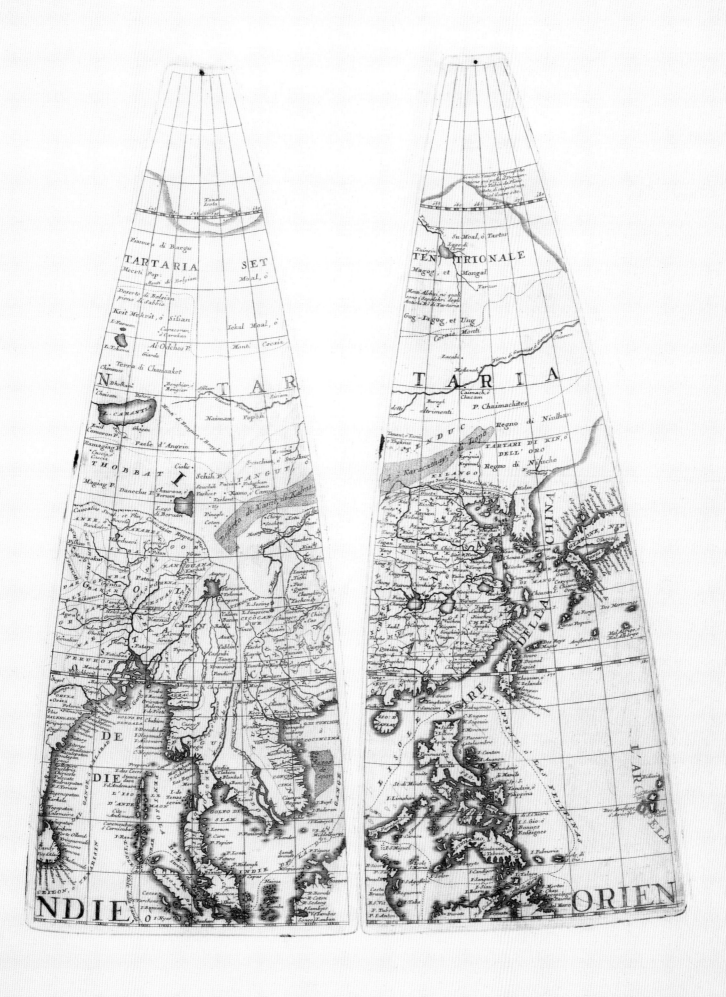

Oculus ♉ Palilicium ♈om.
Aldeboran, regia Cor ♉

Regia card

TAVRVS

Algebar
Rigel. Algeuze

FLVMEN Eridanus
quibufdam N.l.
nov

# THE PARTS OF A GLOBE

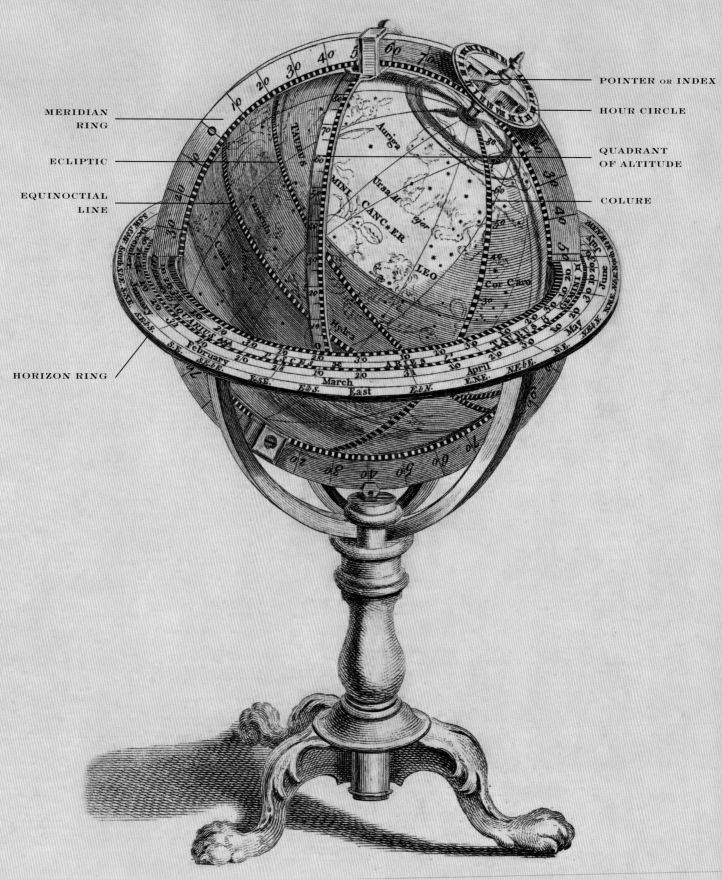

MERIDIAN RING

ECLIPTIC

EQUINOCTIAL LINE

HORIZON RING

POINTER OR INDEX

HOUR CIRCLE

QUADRANT OF ALTITUDE

COLURE

CELESTIAL

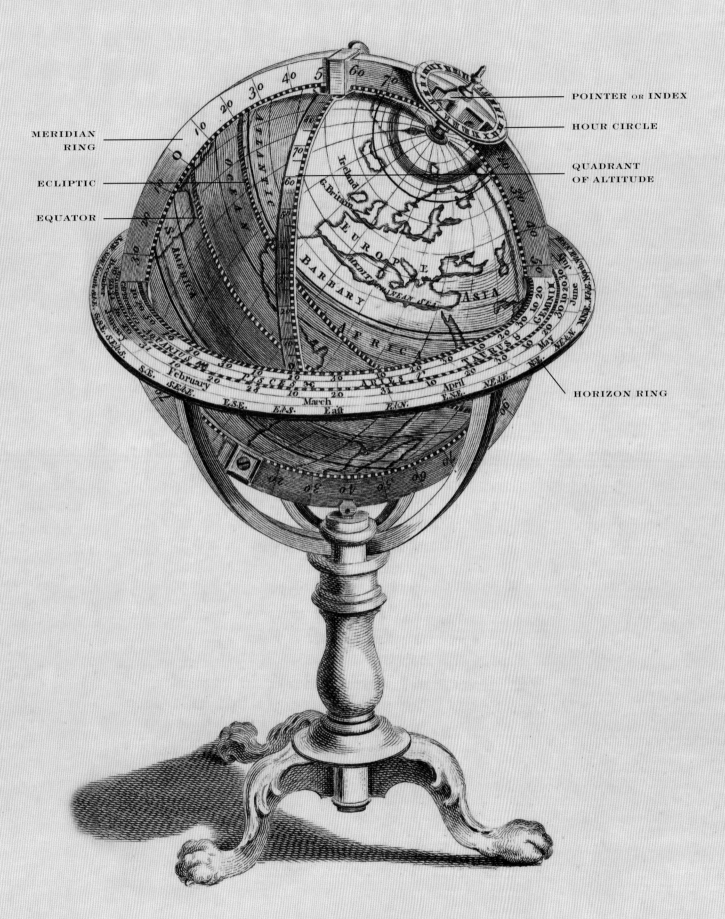

MERIDIAN
RING

ECLIPTIC

EQUATOR

POINTER or INDEX

HOUR CIRCLE

QUADRANT
OF ALTITUDE

HORIZON RING

**TERRESTRIAL**

# A BRIEF HISTORY OF GLOBES

Globes today are spherical maps of the earth or the heavens, and their surfaces, full of detail and colour, always draw our interest and hold our attention. This book shows that in the past they were much more than this. The engraved brass meridian rings and printed horizon rings encircling old globes were not merely decorative features. They were there for a purpose. A terrestrial globe represented the earth and a celestial globe represented the heavens. Together, as a pair, they allowed the relationship between the heavens and the earth to be explored and demonstrated. Globes were more complex objects than they initially appear, and they could be used to perform many tasks. They played a significant role in the distribution of new knowledge and they encapsulated the need to find our place in the cosmos.

The printed globe, as we know it today, emerged in the early sixteenth century. Around this time several factors came together to make it possible and timely for globes to be produced in greater numbers on a commercial basis. We know that globes – terrestrial and celestial – were made before 1500, but very few have survived. Those we know about were one-off items, made for a specific purpose or person, and they were few and far between.

The idea of making spherical models of the earth and heavens originated with the ancient Greeks. It seems the concept of the earth as a sphere was first postulated around the time of Pythagoras in the sixth century BC. This idea gradually came to be accepted by other thinkers of the ancient world. Plato (c. 429–347 BC) alludes to the earth being like a leather ball in his *Phaedo*

of 380 BC, and Aristotle (384–322 BC) too was convinced that the earth was spherical. The first record of an actual globe being made is found in verse written by the poet Aratus of Soli (c. 315–240/239 BC), who describes a celestial sphere with the stars drawn on it made by Eudoxus of Cnidus (c. 408–355 BC). Plato, Aristotle and Eudoxus all theorised about an earth-centred universe. Crates of Mallos (active around 150 BC) is known for the construction of a large terrestrial globe that showed the earth divided into four distinct habitable regions. There are other references to globes being used in teaching, but these very early globes have not survived and it is not known how they were made or what they looked like.

The interest shown by the thinkers of classical times in both the world around them and the visible heavens culminated in the work of Claudius Ptolemy of Alexandria (c. AD 90–c. 168). His principal works – the *Geographia*, a description of the known world, and the *Almagest*, a treatise on astronomy – were of great importance and influence, because for the first time spherical co-ordinates were used to locate the position of places and stars. The earliest extant globe, dating from around AD 150, is the well-known Farnese Atlas, located in the Archaeological Museum in Naples. Featuring a celestial globe, this carved marble sculpture depicts Atlas carrying the weight of the heavens on his shoulders. There are two other celestial globes dating from the same general period. Both are chased (that is, the design is punched and hammered) metal globes. One of them, a brass globe much smaller in size (11 centimetres/4¼ inches), is known as the Mainz globe, because it belongs to the Romano-Germanic Central Museum, Mainz. The other (6.3 centimetres/2½ inches) is made from silver. Known as the Kugel globe, it is in private ownership. After this time, knowledge of the ancients declined in the Christian world and there are very few references to the making of globes for a thousand years.

Fig. 1
**Farnese Atlas, *c.* AD 150.**
Archaeological Museum, Naples.

13

Outside Christian Europe, celestial globes emerged in the Islamic world in the ninth century. They usually took the form of engraved hollow metal spheres. Although they have little relevance to the type of globe explored in this book, as they were not widely known about in Europe, they are a reminder that it was Arabic, Persian and Indian astronomers and mathematicians who developed the work of Ptolemy, until an interest in technical astronomy was rekindled in the West.

In the West, sparse references to celestial globes resurface from the tenth century and one purchased by Nicholas of Cusa (1401–1464) in 1444 survives in Bernkastel-Kues, Germany. There are records of terrestrial globes made in the fifteenth century but the earliest extant example, from 1492 (Globe 1, p. 42), was by Martin Behaim (1459–1507) of Nuremberg, a cultural centre for scholars, artists and instrument-makers, and a crossroads for all types of trade and commerce.

Globes made before 1500 were all manuscript globes – that is, they were unique items, drawn, painted or engraved by hand. They were slow and costly to make and could not be easily replicated. The Behaim globe was an expensive one-off, but unlike previous globes, it had a public impact. The costs of producing the globe were met by the Nuremberg city council and it was displayed in the Town Hall for many years after its completion. It seems likely that the public nature of this ambitious project

to represent the whole world in its proper shape created an influential precedent for subsequent globe-making.

After 1500 globe-making changed dramatically, and the idea of making models of the earth and the heavens for a wider public gained momentum for several reasons. The technical innovation of printing with moveable type in the 1450s had enabled the wider circulation of all types of text and of knowledge in general. Interest in the classical world had been rekindled in the Renaissance; old knowledge was looked at afresh, and new ways of thinking and seeing flourished. At the end of the fifteenth century and in the early years of the sixteenth, some seminal texts, such as Latin translations of Ptolemy's *Geographia* and *Almagest*, were printed for the first time and so became more accessible. These comprehensive books were of particular interest to scholars and stimulated an intellectual and scientific curiosity to know more about the world and the cosmos.

Furthermore, the age of exploration was truly under way. Navigation became a crucial skill for the advancement of trade and conquest, and constantly expanding geographical and astronomical knowledge fuelled the demand for maps and sea charts. By the end of the fifteenth century printing had begun to be used for map-making, and therefore the availability and circulation of maps was greatly increased. Map-makers became concerned with new ways of representing the constantly changing view of the earth and heavens.

*below*
Fig. 5
**The Ambassadors,**
Hans Holbein the Younger, 1533.
National Gallery, London

*opposite below*
Fig. 6
**Still Life with Globe and Cockatoo,**
Pieter Boel, c.1658.
Akademie der Bildenden Künste,
Vienna.

The growth of interest in exploring unknown lands inspired experimentation in the art of globe-making, and the first half of the sixteenth century saw the production of several models, both hand-painted and printed. Printing made it possible to produce globes in greater numbers at lower cost so they could be more widely distributed. The printed globe, terrestrial and celestial, soon became established as the standard type of globe, sometimes called the 'common' globe, and the methods of manufacture changed surprisingly little from the mid-sixteenth century until the twentieth century. Though one-off manuscript globes continued to be made as special commissions for wealthy patrons, this book will concentrate on printed globes produced after 1500 and trace their development through to the start of the twentieth century.

Following Behaim's famous model, the next significant terrestrial globe is that designed in 1507 by the German cartographer Martin Waldseemüller (1470–1521), who worked in Saint Dié, France (Globe 2, p. 45). He is thought to have been the author of the first printed globe gores, a 'gore' being one of the printed segments of paper that is pasted onto the globe sphere (see pp. 33–9 for a detailed explanation of the globe-making process). Waldseemüller is also remembered for giving the name 'America' to the New World (after the explorer Amerigo Vespucci).

The first globe-maker to bring printed terrestrial and celestial globes together as a matched pair was Johann Schöner (1477–1547), a German cleric turned printer who worked in Bamberg and then Nuremberg (Globe 3, p. 46). He made several manuscript globes and started to print them on a commercial basis in 1515.

While terrestrial globes showed the land and sea, celestial globes displayed the layout of the stars. On a practical level the mapping of the heavens was crucially important for seafaring, as sailors navigated across the seas using the positions of the stars. These factors, combined with the contemporary intellectual interest in cosmography (the study and representation of the relationship between the earth and the heavens), made Schöner's pairing of a terrestrial with a celestial globe immensely significant; though each type of globe had a value on its own, together they formed a compact model of the cosmos which would be adopted by globe-makers throughout the next 400 years.

Terrestrial and celestial globes have several elements in common. The framework for both consists of a set of lines: the equator, the ecliptic, the circles of the Tropics, and the Arctic and Antarctic circles. These lines on a terrestrial globe have counterparts in equivalent positions on a celestial globe. In addition, a celestial globe often displays meridian lines called colures, which cross the equinoctial and solstitial points (see the glossary, p. 217, for further explanation of these terms). The user of a celestial globe must imagine the earth at the centre of the sphere and the viewer beyond the heavens, looking down on the universe. The ancient Greek astronomer Hipparchus (active 161–121 BC) established the principle that, whereas star charts, drawn from a viewpoint on earth, show frontal representations of the figures of the constellations, on globes they would have to turn their backs to the user, corresponding to the reversed orientations of the stars.

Waldseemüller had published an explanatory text to accompany his globe, and Schöner relied heavily on this in a manual he published to accompany his own terrestrial globe. Schöner's *Luculentissima quaedam terrae totius descriptio* (*A Very Clear Description of the Whole Earth*) was issued in 1515 and contained additions about new Portuguese discoveries. A manual for use with his celestial globe – *Solidi et sphaerici corporis sive globi astronomici canones usum* (*Manual for the Use of the Solid Spherical Body and Astronomical Globe*) – was published in 1517. Schöner's printed globes, like Waldseemüller's, were made using the technique of woodcut, and for any new edition, entirely new woodblocks had to be made, because the extent to which existing blocks could be altered was limited. Around 1533 Schöner produced a new pair of printed globes, with updated cartography; America is shown on the terrestrial globe, supposedly linked to Asia. His globes became well known throughout Europe. In Hans Holbein's painting known as *The Ambassadors* (1533), the celestial globe that is so clearly depicted on the top shelf is undoubtedly based on one made by Schöner. The terrestrial globe on the lower shelf bears similarities to Schöner's terrestrial globes though no firm attribution has yet been made. The inclusion of globes in the painting illustrates not only the renown of Schöner's work, but also the use of globes as symbols, a constant theme in their history. Globes can signify wealth and power, indicate knowledge and learning, and allude to the temporal or the eternal, depending on the context. In paintings, terrestrial globes often appear in portraits of royalty and state officials to signify authority and possession of foreign lands.

They also appear in portraits of explorers and navigators to signal the nature of their accomplishments. Terrestrial globes in *vanitas* still-life paintings remind us that earthly pleasures are fleeting and finite. A celestial globe is an obvious attribute of an astronomer and they often appear in paintings about alchemical activity, notably in the seventeenth century. Since Roman times globes have frequently appeared on coins and commemorative medals; and when used today in the logos of large companies and institutions, they instantly and obviously lay claim to wide influence and authority.

Though there were other globe-makers in the sixteenth century, for example Georg Hartmann (1489–1564) of Nuremberg and Kasper Vopel of Cologne (1511–1561), Schöner is the best known, perhaps because his globes were more widely distributed. His influence extended well beyond globe-making, as he played a significant role in encouraging the publication in 1543 of *De revolutionibus orbium coelestium* (*On the Revolutions of the Heavenly Spheres*), the treatise by Nicolas Copernicus (1473–1543) that placed the sun, rather than the earth, at the centre of the universe.

The cartography of early globes was based on a mixture of sources. For terrestrial globes information came from the geography of Ptolemy, historic travellers' accounts such as that by the Venetian Marco Polo (1254–1324), charts used by Mediterranean sailors, and the most recently available reports, navigational charts and maps from Portuguese and Spanish explorers. Unknown parts of the world were often imagined, and Terra Australis – the hypothetical large southern continent, thought to balance the northern landmass – continued to appear on some globes until the eighteenth century. Globes became steadily more accurate as detailed information became available. On celestial globes, the forty-eight constellations described by Ptolemy provided the basic design and, as knowledge of the heavens increased and updated star catalogues were compiled, new constellations and stars were added. The two star charts made in 1515 by Albrecht Dürer (1471–1528), Conrad Heinfogel (d. 1517) and Johann Stabius (*c.* 1460–1522), showing the northern and southern celestial hemispheres, were the first printed charts to depict the figures of the constellations. They influenced many subsequent celestial globes, though makers added their own refinements.

Celestial globes were also updated to account for the precession of the equinoxes first quantified by Hipparchus. The equinoctial points (the points at which the equator intersects with the ecliptic, the circle marking the apparent path of the sun through the heavens) shift by one degree every seventy-six years, equivalent to an observed slow change in the orientation of the heavens relative to the earth. The date, or epoch, for which a globe is accurate is often stated on the cartouche, but this is not necessarily the date when the globe was made. Globe-makers usually chose an epoch for a celestial globe close in date to its manufacture.

EFFIGIES GEMMÆ FRISII OBIIT ANNO 1555 Ætatis 47

Schöner had created a market for globes, and as the demand for them increased, others started to take an interest in their production, which involved the skills of many. Roeland Bollaert, a bookseller in the thriving Flemish town of Antwerp, had a keen interest in cosmography and spurred the foundation of a new centre of globe-making in northern Europe. In 1527 he financed the printing of Schöner's manuals in Antwerp, and found people with the practical and intellectual skills needed to make globes in the university town of Louvain. The globes that were produced there proved to be highly influential.

Gemma Frisius (1508–1555) – who was a student in Louvain, where he became a professor of medicine while maintaining an interest in mathematics and astronomy – was commissioned to make a terrestrial globe which appeared around 1530. In order to explain the full use of globes, it became common practice for globe-makers to publish an accompanying manual or treatise. In Gemma's manual, *De principiis astronomiae et cosmosgraphiae deque usu globi* (*Principles of Astronomy and Cosmography and the Use of the Globe*), while acknowledging his debt to Johann Schöner, he called his globe a 'cosmographic globe' because he added several features from a celestial globe. In addition to the ecliptic circle (which appeared on Schöner's terrestrial globes) he added a number of stars. From illustrations in his book it appears that he also adopted the hour circle from Schöner's celestial globe. As with Schöner, Gemma mounted it in a meridian ring set in a horizon ring, which enabled it to be used to solve astronomical

*opposite*
Fig. 7

**Portrait of Gemma Frisius with
globe and other mathematical
instruments, 1557.**
Engraving by Jan van Stalburch.
British Museum, London

*below*
Fig. 8

**Woodcut of the Northern Stars,
Albrecht Dürer, 1515.**
British Library, London.

problems as well as geographical ones. Unfortunately no examples survive, so apart from the illustrations in his manual, we do not know what this globe looked like.

Several years later Gemma published his own pair of globes, only two examples of which are known to exist: a terrestrial globe of 1536 and a celestial globe of 1537 (Globes 4 and 5, pp. 50, 53). Unfortunately their original stands have not survived so we do not know exactly how they were mounted, though an illustration in Gemma's treatise of 1530 depicts a globe in a stand similar to that of Schöner's. They are important as they are the earliest extant printed globes using the technique of engraving. Engraving came to be the printing technique used until well into the nineteenth century, as much greater detail could be achieved than with woodcut. This, combined with their construction (a hollow paper sphere coated with plaster), makes Gemma's globes prototypes for what came to be the standard method of manufacture.

The overall plan for the globes was due to Gemma, but he gathered together information from existing sources and used the practical skills of Gaspard van der Heyden (c. 1496–c. 1549) and Gerard Mercator (1512–1594) to engrave the gores, demonstrating that globe-making was often a collaborative business involving several people. Globe-makers usually started out in another related trade. They could be map-makers, printers or instrument-makers, or have a combination of these skills, and their globe production was carried out in conjunction with these other activities. Family firms were common, with the business often passing from father to son, and people with the necessary skills may have worked with more than one globe-maker.

Gerard Mercator certainly fitted this profile. Today he is probably best known for his maps, and notably his large world map of 1569 based on what came to be known as the 'Mercator Projection'. A few years after his collaboration with Gemma Frisius, he decided to make his own terrestrial globe. It was published in 1541 and had a diameter of 42 centimetres (16½ inches). Mercator wanted his new globe to introduce recent geographical discoveries, especially in southern Asia, but its truly innovative feature was the introduction of rhumb lines or loxodromes (Globe 6, p. 54). A rhumb line crosses all meridians at the same angle, and therefore allows sailors to plot a constant compass direction, or line of the same course, towards the intended destination. Useful tools for navigation, they were first used on portolan charts, the oldest sea charts, used by Mediterranean sailors from the thirteenth to the sixteenth century. On these charts, which did not follow a geometrical projection, rhumbs were drawn as straight lines. On a globe these lines spiral towards the poles. In 1551 Mercator published a companion celestial globe (Globe 7, p. 59).

Mercator's globe profited from the several design improvements made since 1500, to which he added his own refinements. As with earlier models, a Mercator globe was suspended in a brass meridian ring graduated with latitude markings that could be seen easily. The globe, in its brass ring, was mounted in a wooden stand with four legs supporting a printed horizon ring. The basic information printed on this ring comprised a zodiacal calendar, relating the date to the sun's position in the zodiac, and the points and degrees of the compass. Other information – for example, saints' days and the names of winds – was occasionally added. An adjustable graduated quadrant arm (or quadrant of altitude) could be moved along the meridian ring and used for taking angular measurements, while a compass was sometimes inset in the circular base. This design allowed the globe to be used to solve problems relating to the position of the sun and stars at different times of the year. The globe became an analogue computer of the movement of the sun and stars in relation to the earth.

The fifty years following Behaim's globe were a time of significant experimentation and development in globe-making. By the 1540s a model had emerged, in Mercator's globes, which formed the basis for globe production over the next 350 years.

It should be remembered that these early globes were made according to a Ptolemaic view of the cosmos, where the earth was believed to be static at the centre of the universe. The mounting of a terrestrial globe which turned in a stand and rotated about its own axis did not represent the rotation of the earth; rather, it allowed computations to be made relating to time and seasonal change. The user could discover, for example, when the sun rises at a particular time of the year, at a given latitude, which was important in daily life. When the Copernican worldview – where the earth spins about its own polar axis and moves around the sun at the centre of the universe – became the norm, there was no need for terrestrial and celestial globes to change, because their value and usefulness as problem solvers remained the same. From the mid-seventeenth century, models in the form of planetaria and orreries demonstrated the Copernican solar system.

Other printed globes were made in the mid-sixteenth century – for example, those by François Demongenet (active 1550–60) in France (Globes 8 and 9, p. 62) and, around 1570, those by the brothers Livio and Giulio Sanuto (active 1540–88) in Venice – but Mercator dominated the European market in globes until the last decade of the century. His output was prolific and, compared with earlier globes, a fair number have survived. Mercator's globes are considered to be the most important of the sixteenth century.

ÆTATIS — Mercator, tractusque nouos, terræque, marisque — SVÆ. LXII.

Magna tibi priscum tandem superasse laborem,

Monstrasse, et magnum quod continet omnia cælum

Magna Pelusiacis debetur gratia chartis:

I. Vrient. sudb.

Polus magneti

AMERICA

Petru

GERARDI MERCATORIS RVPELMVNDANI EFFIGIEM ANNOR.
DVORVM ET SEX — AGINTA, SVI ERGA IPSVM STVDII
CAVSA DEPINGI CVRABAT FRANC. HOG. CIƆ. IƆ. LXXIV.

*left*
Fig. 10
**Portrait of Mercator and Hondius, from Mercator's *Atlas*, edited by Hondius, Amsterdam, 1619.**
British Library, London.

*opposite*
Fig. 11
**Frontispiece from Joseph Moxon, *A Tutor to Astronomy and Geography*, 1659.**
British Library, London.

A fresh phase of globe-making started around 1586 when a new pair of globes with a diameter of about 33 centimetres (13 inches) appeared in Amsterdam, made by Jacob Floris van Langren (*c.* 1525–1610). Why these globes came to be made is not clear, but seafaring trade had recently become important in the northern provinces of Holland and Zeeland in the Low Countries, and the idea of making globes there, given their link to the art of navigation, is not entirely surprising. Jacob Floris van Langren had also been a pupil of the renowned Danish astronomer Tycho Brahe (1546–1601), which may have inspired him to venture into globe-making. The terrestrial globe was largely based on Mercator's work (especially his wall map of 1569) and Spanish and Portuguese charts. Another source was Lucas Jansz Waghenaer's *Speighel der Zeevaerdt* (*The Mariner's Mirror*) of 1584, a pilot book for the coasts of Western Europe. Van Langren, like Mercator, also included rhumb lines on his globe. Despite battles with upstart competitors in Amsterdam, this family firm continued to make globes until the mid-seventeenth century.

In England, the making of printed globes began around 1592, when Emery Molyneux (d. 1598) produced a large pair of globes measuring 63.5 centimetres (25 inches) in diameter. They were financed by William Sanderson (*c.* 1548–1638), a wealthy merchant, who spent over £1,000 (then a vast sum of money) on

their production. The terrestrial globe celebrated recent English maritime ventures – it depicted the voyages undertaken by Sir Francis Drake (1540–1596) and Thomas Cavendish (1555–1592) and other notable English explorers – while the celestial globe was based on Van Langren's globe (Globes 10 and 11, pp. 65, 70).

Not a great deal is known about Emery Molyneux. He was an Englishman who lived and worked in Lambeth, London, and he was known as a mathematician and instrument-maker. The gores were engraved by Jodocus Hondius the Elder (1563–1612) who, in 1583, had fled to London from Ghent to escape religious persecution. The globes became famous in their day after Molyneux presented a terrestrial globe to Elizabeth I at Greenwich, in a symbolic gesture recognising Britain's increasing world power.

For reasons that are not clear, Molyneux emigrated to Amsterdam in 1597, and he died a year later. Hondius had already moved to Amsterdam in 1593 to set up a globe- and map-making business, much to the anger of the Van Langrens. His competition initiated a period of 'globe wars' in Amsterdam, fuelled by the arrival of Willem Jansz Blaeu (1571–1638) from Alkmaar in 1598.

The output of globes between 1597 and 1605 was truly extraordinary. In these few years seventeen editions of globes

were placed on the market, ranging in size from 10 to 36 centimetres (4 to 14¼ inches). Explorers had recently brought back to Europe much new information about southern lands and stars, and the Dutch East India Company was founded in 1602 to establish trading outposts in Asia. The family firms of Van Langren, Hondius and Blaeu, along with their lesser-known rivals, competed fiercely with each other to bring out globes with the most up-to-date information, in order to have a commercial edge.

By around 1605, Blaeu had become Hondius's main rival. By the end of the decade the fast and furious pace of globe production had slowed a little, but it continued throughout the century. In 1617 Blaeu published a pair of globes with a diameter of 68 centimetres (26¾ inches). They were larger than any other globes made until that time, and went through several editions with revisions. After Willem Blaeu's death, his son Joan Blaeu (c. 1598–1673) took over the business. The copper plates for the globes were eventually sold outside the family and passed through several hands, and the globes continued to be produced until the end of the century, when the final edition was issued in 1696 by Jacques de la Faille (1668–c. 1719), the last person to acquire the plates. By then the globes had become out of date. A considerable number of Blaeu globes have survived,

which indicates that his output was vast, though we do not know the precise numbers of globes made (Globes 12 and 13, pp. 74, 79).

Though globes were often carried by the ships of the Dutch East India Company, it seems the main market for them was an expanding class of wealthy people. Globes were now sold as handsome objects of status and prestige to a comfortable merchant class, in addition to being instruments of geography and astronomy, or navigational and educational aids. Dutch globe-makers were dominant in the first half of the seventeenth century, and their globes were sold throughout Europe, but in the second half of the century, the trade expanded.

Joseph Moxon (1627–1691) became the second person to make printed globes in England and was the first in a new and very productive period of activity. He had spent a great part of his early life in the Netherlands, where his father James, a Puritan, had settled for religious reasons. Following in his father's footsteps, Moxon had learned the printing trade there. After moving back to London around 1650 he set up a printing business and, in addition to books, he printed paper scientific instruments, maps and, most importantly, globes. Moxon's first pair, with a diameter of 35.5 centimetres (14 inches), was published in 1653. In 1654 he published his first book on the subject of globes, which was actually a translation of Willem Blaeu's 1634 manual on globes, *Institutio astronomica de usu globorum et sphaerarum* (*An Education in Astronomy and the Use of the Globes and Spheres*). Moxon's new globe manual, *A Tutor to Astronomy and Geography; Or an Easie and Speedy Way to Understand the Use of Both the Globes*, was published in 1659. It was written specifically for the English public, with London being the main reference point rather than Amsterdam, as it had been in Blaeu's manual and Moxon's translation.

An example of a problem given in the manual to be solved using the globes is as follows:

## PROB. XXII.
### *The place of the Sun, and Hour of the Day given; to find its Azimuth in any assigned Latitude.*

The Globe,&c. Rectified to your Latitude; Turn the Globe till the Index of the Hour-Circle come to the given hour; and bring the Quadrant of Altitude to the place of the Sun; fo shall the number of degrees contained between the East point of the Horizon, and the degree cut by the Quadrant of Altitude on the Horizon, be the number of degrees of the Sun's Azimuth, at that time.

Moxon's globes and books were immensely successful and ran to several editions in his lifetime. He advertised globes in several sizes, ranging from 3 to 26 inches (7.5 to 66 centimetres). The diarist Samuel Pepys (1633–1703) acquired a pair for himself and also ordered a pair for the Admiralty during his time as Chief

Secretary. Moxon is often described as the inventor of the pocket globe, though there are Dutch claims to this title. The pocket globe was a mini-cosmos: a small terrestrial globe enclosed in a fish-skin case lined with celestial gores (Globe 16, pp. 89).

Following Moxon's success, other globe-makers started to emerge in London. An apprentice to Moxon, William Berry (1639–1718) joined forces with Robert Morden (active 1650–1717), another map- and instrument-maker. In 1675 they advertised a new treatise on the use of globes and a proposal for a new large globe, 30 inches (76 centimetres) in diameter, which would be more up to date than any globes yet made in England or the Netherlands. No Berry and Morden globe of this size survives, and it may be that none was ever made; but several smaller globes with a diameter of 35.5 centimetres (14 inches) still exist. On the smaller globes an address to the reader placed in a cartouche in the Indian Ocean states 'Indeed there is not any part of ye earth wherein we have not made a considerable alteration' to indicate that their globes displayed new and more accurate information.

Around the same time, the Venetian Franciscan monk Vincenzo Coronelli (1650–1718) was also making globes. He started off with manuscript globes, and the enormous pair of painted globes he made for King Louis XIV of France, between 1681 and 1683, brought him international recognition (Figures 13, 14). These globes, known as the Marly globes after the palace of Marly, where they were first displayed, still exist and have a diameter of 3.85 metres (12 feet 7½ inches). In 1686 Coronelli founded a workshop for making printed globes at the convent of Santa Maria Gloriosa dei Frari in Venice.

Coronelli liked to work on a grand scale. The first pair of printed globes he published were 108 centimetres (42½ inches) in diameter (Globes 17 and 18, pp. 90, 93). They appeared in 1688 and were followed by several later editions. With his celestial globe gores he did something unusual: in addition to the normal convex globe gores one finds on globes, Coronelli printed a set of concave gores to show the heavens as we actually see them. In order to work, they would of course have to be pasted on the inside of a sphere, which is slightly impractical. Coronelli's globes can be found in grand settings all over Europe. His output was prolific and he also made globes in several smaller sizes. As part of a series of geographical atlases, he published a collection of his gores in book form in the *Libro dei Globi* (Globe 19, p. 96).

Throughout the seventeenth century there were other people making globes, inspired by the work of their predecessors. In Rome, for example, in the 1630s Matthäus Greuter (*c.* 1556–1638), originally from Strasbourg, made globes of two sizes, 26.5 and 49 centimetres (10½ and 19¼ inches). The terrestrial globes were based on the large globe of Willem Blaeu and the celestial on a globe by Pieter van den Keere (1571–*c.* 1646) and Petrus Plancius (1552–1622), who are better known for their maps than their globes. Greuter's globes were reissued later in the century by Giovanni Battista de Rossi (active 1640–72). Another member of the family, Giuseppe de Rossi, had already published copies of globes by Jodocus Hondius in 1615. Isaac Habrecht II (1589–1633), a mathematician and physician in Strasbourg, published a small pair of globes in 1621 and Jean Boisseau (active 1631–48) in Paris copied globes by Pieter van den Keere. Other isolated forays into globe-making are known: in Paris, Pierre du Val (1619–1682), an engraver, produced a set of gores around 1666.

The founding of the Royal Society in London in 1660 and its equivalent in Paris, the Académie Royale des Sciences in 1666, heralded a more formal and structured approach to scientific enquiry and the investigation of the natural world, which continued throughout the eighteenth century, the Age of Enlightenment. There were also advances in the field of astronomy. From 1609 the telescope was used to examine the heavens. More stars were observed and new catalogues were made:

these were major undertakings and two important examples were posthumously published, by Johannes Hevelius (1611–1687) in 1690 and by John Flamsteed (1646–1719) in 1725.

In France, one of the members of the Académie, Giovanni Domenico Cassini (1625–1712), Director of the Paris Observatory, had a large influence on its geographical and astronomical research. Throughout the eighteenth century, expeditions to distant parts of the world returned with new information that was used extensively in map- and globe-making. Guillaume Delisle (1675–1726), a pupil of Cassini and member of the Académie from 1702, put fresh data to use in a pair of globes in 1700. He used the latest information to plot his globes, eliminated errors that had been perpetuated on earlier examples, and left uncharted territory blank, elevating French globes to the status of serious scientific instruments.

Although other globe-makers started to emerge – for example Nicolas Bion (1652–1733), Louis-Charles Desnos (1725–1791) and Jean-Antoine Nollet (1700–1770) – the overall production in France was small, and it was not until later in the century that globe-making became firmly established there. Two outstanding French names from this period are Didier Robert de Vaugondy (1723–1786) and Charles-François Delamarche (1740–1817, Globe 46, p. 185).

In London, globe-making was already well established in the early eighteenth century. Hermann Moll (1654–1732, Globe 21,

p. 106) Charles Price (*c.* 1679–1733) and Richard Cushee (1696–1733, Globes 26–8, pp. 121, 122, 126) all made globes at this time, but in the first part of the century it was John Senex (1648–1740) who dominated the field (Globe 25, p. 118). He was highly respected, and after his death his widow Mary took over the business. Many years later, James Ferguson (1710–1776) and then Benjamin Martin (1705–1782) acquired the printing plates for Senex's globes and started publishing their own globes. In recognition of his standing they continued to cite his name on their own output (Globes 29 and 30, pp. 129, 132).

As in France, globes in England were bound up with the pursuit of scientific knowledge and were regarded as scientific instruments as well as cartographical and navigational tools. In the latter half of the century, George Adams (1709–1772), a renowned instrument-maker, came to dominate the globe trade. He became mathematical instrument-maker to the Office of Ordnance in 1748, and in this role he supplied many types of instrument to the government. In 1760 Adams was appointed mathematical instrument-maker to George III and in 1769 he was commissioned by the Royal Society to supply the instruments used by James Cook on his expedition to the South Seas to observe the Transit of Venus (Globes 33 and 34, pp. 144, 150).

Many colleges in Oxford and Cambridge have globes by Senex or Adams installed in their libraries. Throughout Europe,

the libraries of scholarly institutions and courts have acquired globes since they started being made. In addition to being functional objects in these settings, they provided a handsome focus of interest. Before the mass production of globes in the nineteenth century, they were relatively expensive for individuals to buy, and therefore owning one, or a pair, could be a status symbol for the owner. The symbolic nature of a globe was reinforced in many ways. They adorned the title pages of atlases, scientific treatises and trade cards; their instantly recognisable form immediately conjures up a connection to learning.

An interest in globe-making was revived in Germany by Johann Baptist Homann (1664–1724), who set up a map-making and publishing house in Nuremberg in 1702 that was run by his heirs into the nineteenth century. Though there is only one globe that bears his name (Globe 22, p. 109), the firm of Homann was associated with several other globe-makers, including Johann Gabriel Doppelmayr (1677–1750), who is one of the best-known from this time (Globes 23 and 24, pp. 110, 114). Though Doppelmayr dominated German globe production at this time, small numbers were made elsewhere. Georg Moritz Lowitz (1722–1774), working for the Homann publishing house, produced a pair of 13.5 centimetres (5¼ inches) in 1747, the cartography of the celestial globe being based on John Flamsteed's star catalogue of 1725.

Fig. 17
**Advertisement for Richard Cushee globes, *c.*1731.**
British Library, London.

ANDERS ÅKERMAN
*Graveur wid Kongl. Wet. Societeten i Upsala.*

After the demise of the Blaeu family enterprise in the Netherlands, completely new and more accurate pairs of globes, based on recent French exploration and using the latest star catalogues, were produced in Amsterdam by Gerard Valk (1652–1726). Between 1700 and 1726, seven sizes of globes were published. Revised editions were brought out mid-century by his son Leonard (1675–1746) and then by Petrus II Schenk (1692–1775), who took over the business after Leonard's death. The production of Valk globes was long-lived, as Cornelis Covens (1764–1825), who came from a well-known map-making family, bought the Valk factory in 1800. He had seen the new globes of George Adams's son, also George (1750–1795, Globe 41, p. 169), and, inspired by them, devised a new mounting for his globes to demonstrate the Copernican movement of the earth.

The abundance of new information coming to light galvanised more globe-making activity in Europe. Sweden had founded its own Royal Academy of Sciences in 1739 and the University of Uppsala founded the Cosmographical Society in 1758 specifically to expand geographical and astronomical knowledge. Accurate globes were necessary for this venture and the Cosmographical Society commissioned Anders Åkerman (c. 1722–1778), an engraver and mathematician, to make the first Swedish globes. They were published in 1759 (Globes 31 and 32,

pp. 136, 141). After Åkerman's death, globe-making was continued in Sweden by his pupil Fredrik Akrel (1748–1804).

At the same time in Austria, Peter Anich (1723–1766) was experimenting with making globes. He took private lessons in mathematics, geography and astronomy and, after the production of a large pair of manuscript globes, he published a 20-centimetre (7¾-inch) printed pair in 1758 and 1759. Anich was unusual in that he lived and worked in a Tyrolean village (rather than a large town or city) and carried out all aspects of production himself. Drawing on contemporary maps, he assembled the cartography for his globes, engraved the gores, made the spheres and turned the stands. He became well known as 'the peasant cartographer' for maps as well as globes. In the larger cities in this part of Europe, Vienna and Prague, globe manufacture on a commercial basis did not establish itself until the 1820s, when Joseph Jüttner (1775–1848) and Franz Lettany (1793–1863), who operated in both these cities, published a terrestrial globe of 31.5 centimetres (12½ inches) in 1822 and a celestial globe two years later. Following the popularity of Jüttner and Lettany's globes, Tranquillo Mollo (1767–1837) published a slightly smaller terrestrial globe of 21 centimetres (8¼ inches) in Vienna in 1824.

The globes of Vincenzo Coronelli did not inspire a continuation of globe manufacture in Italy, and after his death in 1718 little globe-making activity took place there until the

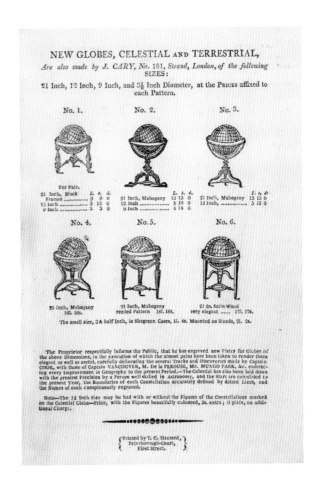

end of the century, when globes started to appear again, most notably in Rome, where the engraver and map-maker Giovanni Maria Cassini (1745–*c.* 1824) published a pair based on the latest discoveries by French and English explorers (Globes 39 and 40, pp. 164, 166).

During the eighteenth century, knowledge of new discoveries and advancements in learning was made more easily available to anyone interested by way of public lectures and the publication of accessible books written with a keen and educated audience in mind. Globes were desirable items in any learned person's library. The rapidly expanding middle classes attached great importance to the education of their children. Geography and astronomy were added to reading, writing, arithmetic and religious studies, the subjects that were considered essential for a proper education for all children (Globe 38, p. 163).

By the end of the eighteenth century, globes were being produced in much greater numbers throughout Europe. Many of the established firms continued their trade into the next century and, seeing a commercial opportunity, several new globe-makers appeared. In addition, the first decade of the nineteenth century saw the first globes to be commercially produced in America.

A terrestrial globe continued to be a perfect vehicle for showing new knowledge about the world in compact form. There was still much uncharted territory to be explored, and

new colonial settlements were being made. Wars were fought and borders redrawn, and more people were travelling further afield. New discoveries and changes needed to be recorded and disseminated. During the course of the nineteenth century, the education of children of all classes broadened and the teaching of geography became much more common. The demand for cheaper globes was satisfied, aided by the introduction of colour lithographic printing, and imaginative ways were found to make geography interesting and fun. Many more toy, puzzle and novelty globes appeared. Globe-making went from strength to strength (Globes 47, 56 and 57, pp. 186, 207).

In England the names Bardin, Cary, Newton, Malby and, later, Wyld dominated the field (Globes 36, 37, 44 and 48, pp. 156, 161, 179, 189). These were family firms, with the business passing from one generation to the next and often involving wider family members. Globe-makers of the nineteenth century were more associated with map-makers, rather than instrument-makers, as had been the case in the previous century. Makers' names appeared on their globes in different formats, depending on the constitution of the firm at any particular time, and their advertisements touted a large range of globes to suit any purse.

Globes continued to be made in pairs, but as time progressed the parity of the celestial globe with the terrestrial globe waned. From the early 1800s, a succession of popular and inexpensive

Fig. 21
**A lesson in geography in Washington, D.C., 1899.**
Photo by Francis Benjamin Johnston. Library of Congress, Washington, D.C.

star atlases and maps was published, which provided cheaper substitutes for celestial globes; and with the advance of astronomy through ever better telescopes, fainter stars could be seen, but it was difficult to show them on a globe. Furthermore, in the nineteenth century the everyday reliance on the heavens for time-keeping purposes diminished with the proliferation of accurate clocks and watches. In 1810 John Cary (1755–1835) advertised celestial globes 'with or without the Figures of the Constellations', charging 5 shillings extra to have the figures 'handsomely coloured'. As the century progressed, the depiction of the constellations became plainer. Though celestial globes continued to be made, their decreasing relevance and their expense rendered them optional rather than necessary.

Elsewhere in the British Isles, after Scotland's first globe – a solitary pocket globe, published in 1793 by John Miller (1746–1815, Globe 42, p. 172) – globe-making eventually became established in Edinburgh with the 30.5-centimetre (12-inch) globes of James Kirkwood (c. 1746–1827) and Alexander Donaldson (active 1799–1828), followed by William Johnston (1802–1871) and his brother Alexander Keith Johnston (1804–1871, Globes 59 and 60, pp. 213, 214).

In Germany, globe-making continued in Nuremberg, with Johann Georg Klinger (1764–1806) setting up a company in the 1790s to cater for an expanding domestic market. The

firm made globes in several sizes, and after Klinger's death his widow ran the business. The firm was taken over by Johann Paul Dreykorn (1805–1875) in 1831 and continued under different owners, but still using the Klinger name, until just after the First World War. Through the century other globe-makers worked for the firm: Carl Abel (active 1852) and Johann Bernard Bauer (1752–1847), for example, who also made globes under their own names. Towards the end of the century, the firm published globes in several languages, in order to capture an international market. Nuremberg, however, relinquished its standing as the centre of globe-making in Germany. The Geographical Institute in Weimar, founded in 1804, enthusiastically promoted the manufacture of globes, and by the 1860s it presented thirty-five different models for sale. Berlin too saw a steady increase in globe production during the course of the nineteenth century, with the names Dietrich Reimer (1818–1899) and Ernst Schotte & Co. rising to prominence (Globe 55, p. 204).

Many more globe-making operations materialised in Europe in the second half of the century. Jan Felkl (1817–1887) founded a company in Prague. Initially his globes were engraved and hand-coloured, but the adaptation of colour print lithography to globe-making allowed them to be produced much more cheaply. The Felkl firm's output was vast, and they eventually supplied globes in a remarkable seventeen languages, catering for the regional differences of the Austro-Hungarian Empire in addition to the rest of Europe.

As in Europe, the teaching of geography and astronomy was part of mainstream education in American schools by the end of the eighteenth century, and maps and globes were promoted as educational aids. Foreign globes were expensive to buy and import, so it is not surprising that a market surfaced on American soil. There are records of globes being made in the late eighteenth century, but the first successful commercial globe-maker in America was James Wilson (1763–1855) of Vermont (Globes 50 and 51, pp. 192, 195). Unlike most of his European counterparts, Wilson came from a farming background. Through dogged determination and by teaching himself all the necessary skills, he started to produce globes commercially around 1810. Other globe-makers soon emerged. Boston became a globe-making hub. William Annin (active 1813–39), an engraver, constructed his own globes, but he also engraved gores for the bookseller Josiah Loring (1775–1840) and Gilman Joslin (1804–c. 1886). Dwight Holbrook (active 1820–30), also based in Boston, made teaching apparatus for schools, including small globes. In New York, Silas Cornell (1789–1864) also successfully made globes.

A set of original gores by Gerard Mercator was discovered in a private library in Belgium in 1868. The discovery was greeted with excitement and the Royal Library in Brussels bought them. The find was marked by the publication of a set of facsimile gores using a new photographic process, and it stimulated a fresh interest in the early days of globe-making, with new scholarly research being undertaken. The globe had come full circle (Globe 58).

We do not know how many globes each maker produced, but the numbers that still survive are testament to a vast output. The use of globes has changed over the years and, apart from simply adding new information, individual globe-makers have always found ways to improve their globes, to make them more special and to differentiate them from their competitors. As we have seen, they were more than three-dimensional geographical and astronomical maps. They became instruments for the theoretical measurement of time. They were demonstrational tools for navigators and a teaching device for the education of children. From their inception, they were beautiful objects to look at and had a decorative function, but were also used as symbols of learning, knowledge, power and status.

Fig. 22
**James Wilson, drawing by Roy Frederic Heinrich, 1810.**
Library of Congress, Washington, D.C.

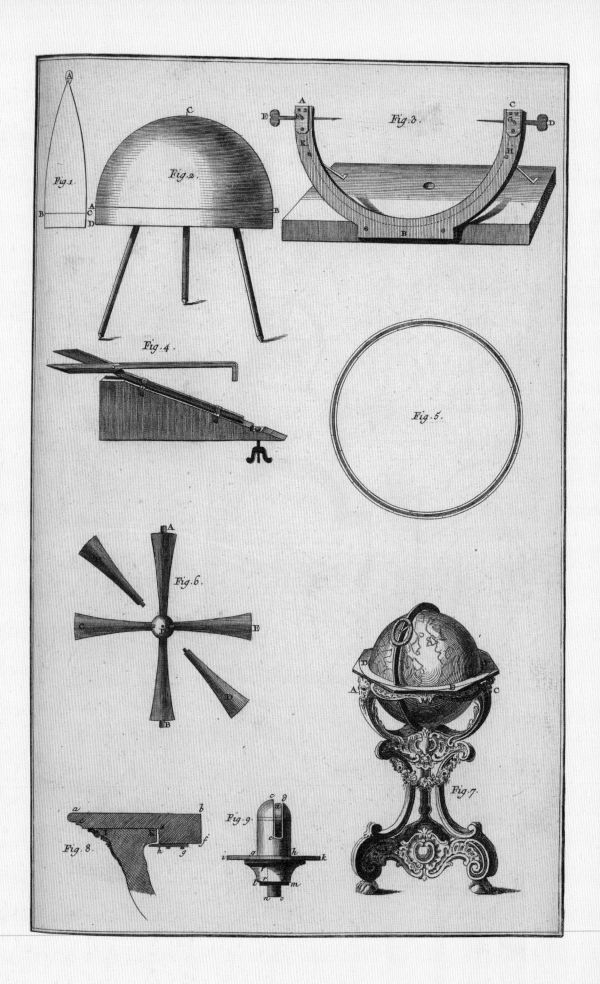

# THE MAKING OF GLOBES

Globes can be made from many materials, but remarkably, for the vast majority of printed globes, the basic method of construction that was developed in the first half of the sixteenth century remained more or less the same until the end of the nineteenth century. The apparent perfection of their smooth surfaces, however, does not immediately reveal the techniques underlying their construction and the difficulties their makers have had to surmount in order to wrap a flat map around a sphere. Globe-makers were not particularly forthcoming about their methods of construction, but a few accounts of the actual process were published. There is considerable variation in the level of detail but there is agreement about the general method and materials of manufacture. Examination of damaged globes during the course of conservation work has confirmed these methods, and has also revealed some secrets.

Edmund Stone described the use and parts of the globe in *The Construction and Principal Uses of Mathematical Instruments* (1723), his translation of Nicolas Bion's work. He also gave a full account of how they are made, explaining that 'the Body of the Globe is composed of an Axle-Tree, two Paper-Caps sewed together, a Composition of Plaister laid over them, and last of all globical Papers or Gores, stuck or glewed on the Plaister.' This is essentially the method of construction used for most printed globes.

One of the most eloquent descriptions of globe-making is that by Didier Robert de Vaugondy, published in Denis Diderot's *Encylopédie des Sciences, des Arts et des Métiers* in 1757. His account, in which he continually stressed the need for excellence and high standards of workmanship, helpfully includes an illustration of the equipment required. Firstly, in order to form a sphere, a mould was needed. Vaugondy described the use of a wooden half-ball, preferably made of hardwood as it was less likely to split. (He particularly recommended using the forked roots of the knotted elm which had been exposed to the sun for a long time.) Edmund Stone differed on this point, suggesting a whole wooden ball for this purpose. For larger globes, wooden moulds are impractical and, as listed in an inventory of equipment that belonged to the Dutch globe-maker Jan Jansz van Ceulen (drawn up in 1689), hollow copper hemispheres were used. They could, of course, also be used for smaller globes. The mould had to be slightly smaller than the diameter of the intended globe to leave enough space for the paper caps and plaster.

Vaugondy's procedure involved first coating the mould with moist soap to prevent the shell from sticking to it, and then forming a paper hemisphere by covering the mould with three overlapping layers of thin, dampened, gore-shaped paper board, glued together with flour paste. After the application of the final layer they were all tied together around the equator and allowed to dry. Vaugondy helpfully added that one can speed the process up by having two moulds, and that summer was the best time of year for making the shells, as they then dry more quickly. Stone differed again. He suggested pasting 'waste paper, both brown and white' over the mould for the sphere. The use of alternate layers of brown and white paper, he argued, was very practical in that it allowed the globe-maker to see when each covering layer of paper has been completed. When the thickness of the paper layers was judged to be rigid enough (in some accounts, about the thickness of a crown coin piece), and before it was quite dry, it was to be cut down the middle at the equator and taken off the wooden mould.

Fig. 23
**Equipment and process of globe-making, from Denis Diderot, *Encyclopédie*, 1757.**
British Library, London.

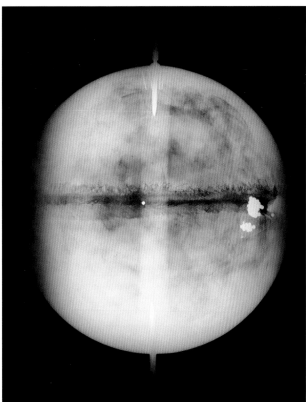

*left*
Fig. 26
**X-radiograph of a celestial globe by Benjamin Martin, after 1757 (30.5 cm/12 in.).**
The nails attaching the shell to the wooden support pillar are visible. The join at the equator is reinforced with a strip of rough cloth. A bag of lead shot, some of which has escaped, is present.

*above*
Fig. 27
**Interior of a celestial globe by Newton & Son, 1851 (30.5 cm/12 in.).**
A bag of lead shot, glued to the shell, has been used to balance the globe.

Globes are often described as having a papier mâché shell. While there is a common assumption that this term refers only to pulped paper, it can also include laminated paper.

For globes, the hemispherical shells were generally formed, as in the descriptions above, from layers of paper or thin board rather than mashed paper pulp. Where damage has allowed a look inside a globe, all kinds of paper can be seen: rough brown paper, book pages, maps, paper scraps with manuscript, and general ephemera.

Once the shells had been removed from the mould, the inner support could be positioned and secured. Occasionally the support (Stone's 'axle-tree') was simply an iron rod, which might be secured internally with wooden blocks at the poles, but the most common internal support was a wooden pillar with pivot pins driven into the ends. The pillar was placed inside the shells, so that the pivot pins went through holes at the north and south poles, and it was secured with glue and small nails. The pillars are usually wider at the ends than at the centre to give a wider area of support at the poles, and to provide enough space for the securing nails. On larger globes, support structures with branches extending from the centre were sometimes used, with the ends of the branches, sometimes two, sometimes four, aligning with the line of the equator. Nails or staples secured them in place. Tell-tale signs of a branched structure are small circular cracks in the shell at intervals of 90 or 180 degrees along the equator.

The two paper hemispheres were then joined together. Stone instructed that this should be done by sewing with strong twine. This is sometimes seen, but often the join was made by applying glue along the edges and sealing it with more paper or a strip of cloth. Though the inner shells of most globes are joined at the equator, they can also be fastened longitudinally so the join passes over the poles. Dutch globes were often made in this way.

The joined hemispheres formed a paper ball, which was then coated in plaster. The ball was placed in a steel semi-circular former having the exact diameter required for the globe, the pivots being secured into notches to keep them steady. Plaster was then applied while the ball was turned. The inner semi-circular edge would remove the excess plaster and eventually a sphere would emerge, as Stone puts it, 'like a Ball of polished Marble'.

The globe was then tested to see if it balanced properly. Ideally the weight of a globe should be distributed evenly over the whole surface, but the layer of plaster may have been slightly uneven in places, causing the globe to swing to one position. This was remedied by cutting a small hole in the shell and strategically gluing a weight to the inner shell to 'balance' the globe. The weight was often simply a cloth bag filled with lead shot. This is usually the cause of a rattling globe: when, at a later stage, the bag splits and the lead shot spills out, or the whole bag falls to the lowest part of the sphere, there is a clunk when the globe is turned.

A globe shell can vary in thickness from about 3 to 10 millimetres (⅛ to ½ inch), and can be unexpectedly thin, even on a large globe. This simple construction had several benefits: it did not warp as wood may have done and, for the profit-minded globe-maker, the raw materials were relatively cheap: old scraps of paper could be recycled. Damage can occur through impact, but repairs are often possible. The plaster provided an ideal surface on which to paste the gores. Guidelines for the gores were sometimes drawn or scored on the plaster surface; however, it is by no means a simple matter to cover a sphere with paper so that the surface is smooth and wrinkle-free. To make it flat, the paper has to be stretched, meaning that any printing on the paper will also be stretched, which had to be taken into account when the gores were designed. Furthermore, to compound the problem, when paste is applied to paper it gets wet and expands. Any inaccuracies would result in a globe with overlapping gores, or worse still, gaps between the gores. On the evidence from the globes themselves, these difficulties were usually overcome.

The existence of the gores made by Martin Waldseemüller in 1507 (Globe 2, p. 45) shows that thought had been given to gore construction early in the sixteenth century. Albrecht Dürer, in his book on measurement, *Underweysung der Messung* (*Instruction on Measurement*), published in 1525, describes a sphere laid out flat, cut along sixteen meridian lines, but the earliest illustrated account specifically referring to globe gore construction appears to be that by Henricus Lorius Glareanus (1488–1563) who, in his small treatise on geography of 1527, provides a diagram showing the construction of twelve gores. Various geometrical solutions were devised over the centuries to compensate for the distortion that occurs when flat paper is transferred to the round. The use of twelve gores came to be standard practice, 360 degrees being conveniently divisible by twelve. On larger globes eighteen or more gores were sometimes used, and in order to make their application to the sphere easier, they were often cut: at the equator to give half-gores, slit partway along the central meridian of each gore, and clipped about 20 degrees from the poles, the polar area being covered by a circular cap, also known as a calotte. On larger globes, the gores were sometimes cut along latitude lines to reduce the size of each section. Polar caps provided a solution to the difficulty of bringing numerous gore tips together neatly at the poles. Though such caps are usually seen on terrestrial globes, they were not always employed on celestial globes. Furthermore, celestial gores can be aligned with the equator or the ecliptic. If aligned with the celestial equator, the gore tips will meet at the celestial poles, and if aligned with the ecliptic, the tips will meet at the ecliptic poles. Ultimately this was a matter of personal preference for each globe-maker. The pivots, in keeping with the apparent rotation of the heavens, are always at the equatorial poles. The alignment of gores with the ecliptic does not mean the user must adopt ecliptic co-ordinates.

As we have seen, the printing technique used for the earliest globes was woodcut, a relief process in which the design was chiselled into a block of wood cut along the grain, and the parts of the block not cut away were inked. By the mid-sixteenth century, engraved globe gores had become the norm. With engraving, an intaglio process, the lines that held the printing ink were incised into copper plates. Even though copper plates were more expensive than wood blocks, engraving had many advantages. It was easier to make fine and curved lines, and to incorporate small lettering so that much more detail could be included. Copper plates could also be altered more easily and re-engraved. Larger plates could be made and, being more robust than wood, they lasted longer, so more impressions could be taken. Globes continued to be engraved until the latter part of the nineteenth century, but from the mid-century, globes started to be printed using colour lithography. Lithography, a printing process which exploits the

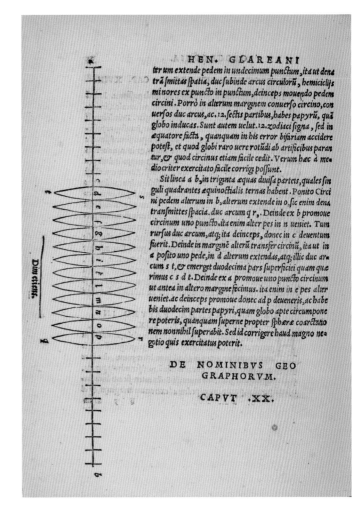

*above*
Fig. 28
**Diagram by Henricus Lorius Glareanus, 1527.**
The first printed diagram showing gore construction.

*opposite left*
Fig. 29
**Page from *Pantologia: A New Cyclopaedia*, volume 5, 1813.**
Illustration showing gore construction.

*opposite right*
Fig. 30
**Page from Denis Diderot, *Encyclopédie*, 1757.**
Illustration showing gore construction.

repulsion of oil and water, and which requires no mechanical treatment of the printing plate, had been invented at the very end of the eighteenth century and was soon adapted for commercial use. The use of lithography marked another change in globe production: they could be made at lower cost and in even greater numbers. Lithography gradually took over from engraving as the main printing technique used for globes.

A wide range of paper was used for the printing of gores. It can be surprisingly thin in some cases, considering the manipulation it had to go through to be stretched and coaxed into place, and then burnished to ease out wrinkles. A starch-based paste was commonly used to attach the gores to the plaster shell. Only then would the gores and horizon ring be coloured. Until the introduction of lithographically printed globes, all colour was applied by hand. Colour, in addition to being decorative and attractive, made information easier to read. Essential lines, such as the equator, became more obvious. Pale washes of transparent watercolour defined areas of land but did not obscure the printed information, while boundaries were delineated with a bolder pigment of the same hue. Likewise, on a celestial globe, colour clarified the constellations. Important features such as stars and towns were sometimes highlighted with spots of gold. Not all globes by the same maker were coloured in the same manner. Purchasers of globes could sometimes pay more to have them elaborately coloured, and a globe commissioned for a high-ranking individual typically has more colour than a standard model. Sometimes colour was added long after the globe was made.

Though it is not clear whether early globes were varnished as a part of their making, it was at some point discovered that a layer of varnish would provide protection against dust, dirt and inquisitive fingers. Globes, after all, were made to be touched and turned. Varnish also improved the aesthetic appearance, making the colours appear richer and brighter. Before the varnish could be applied, the gores had to be coated with a protective layer to prevent the varnish from sinking in and discolouring the paper.

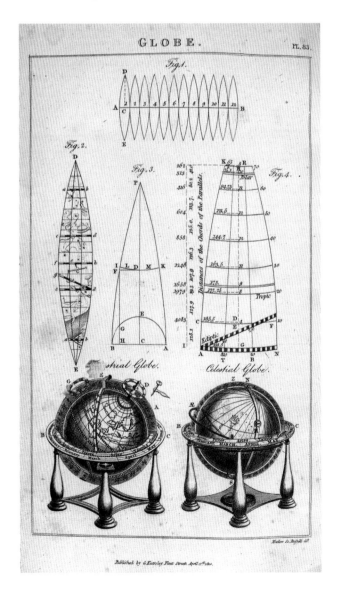

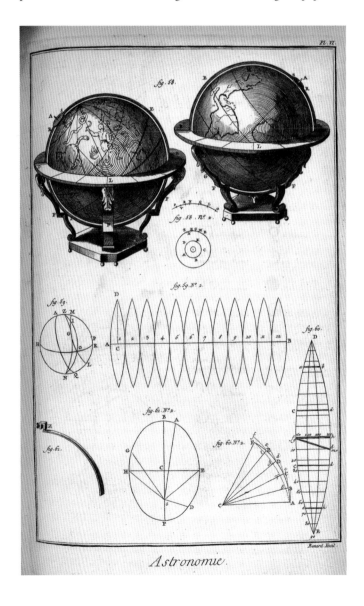

Fig. 31
**A terrestrial globe by
J. & W. Cary, 1839 (46 cm/18 in.).**
The original varnish on the left had
become discoloured, scratched and
dirty. On the right, the varnish has
been removed and the condition of the
underlying paper is revealed. Although
the varnish itself had become damaged,
it had protected the paper well.

Fig.32
**A celestial globe by
Thomas Malby, 1869 (46 cm/18 in.).**
Damage to this globe includes a
split at the equatorial join, and
losses of paper and plaster.
The varnish has become
discoloured by natural ageing.

The paste used to apply the gores to the sphere may have been
adequate, but additional coatings of paste or gelatine may have
been used. When first applied, the varnish, usually a simple spirit
and natural resin varnish, would have been fairly colourless, so the
delicate tinting could be seen. The dark brown appearance of so
many globes today is the result of natural ageing of the varnish.

A globe became complete when it was mounted in its stand,
sometimes called a frame, the primary purpose of which is to
support the globe sphere securely while allowing it to turn freely.
The small globes that survive from the early sixteenth century
were mounted in tripod stands but, as globes became larger,
something more stable was required. Gerard Mercator's globe
stand – a circular base, with four legs supporting the horizon ring,
into which the globe in its meridian ring could be easily fitted
and supported at its base – became a classic design. It was simple,
functional, stable and adaptable. This basic design, sometimes
with adjustments, continued to be used by globe-makers
everywhere until the end of the nineteenth century. For larger
globes, more legs could be added for additional support and,

later on, the circular base was phased out and the flat stretchers
which supported the circular base were replaced with turned
stretchers. It should be noted that there is a correct way to position
a globe in its stand. The graduated face of the brass meridian
ring should be aligned with the north–south axis on the horizon
ring. Placing the globe incorrectly in its stand can cause damage
through abrasion.

The use of oak, which was easily available and durable, was
common in the sixteenth and seventeenth centuries. Different
types of wood, such as ebony or pear wood, could be incorporated
into the design to give contrast and enhance the look of a stand.
Smaller stands were often made with pear wood, which has a
finer grain than oak and is easier to work. Gradually, the design
of stands became more varied, as tastes in furniture changed.
Globe stands started to reflect the availability of more exotic
woods brought to Europe by foreign trade. In England, stands
in mahogany and satinwood were being advertised at the end
of the eighteenth century, in addition to cheaper stands, simply
described as 'stained'. The tripod stand, in wood, made

a comeback in the late eighteenth century, and single-pillar stands in the nineteenth century satisfied the demand for cheaper globes. To protect globes, leather covers that slipped over the upper hemispheres and horizon rings were sometimes made, though few survive.

The construction of pocket globes was divided into two types. They could be solid wooden spheres or miniature versions of the standard globe. The cases for pocket globes were made in several ways. Often they comprised thin strips of wood moulded into a hemispherical shape and glued together, coated on the inside with a thin layer of plaster and covered on the outside with sharkskin or leather.

The paper and plaster shell was a simple but ingenious way of constructing globes of all sizes, and as we have seen, if treated well, it can be surprisingly robust, durable and stable. This technique of production continued well into the twentieth century but other mass-manufacturing processes, and materials such as tin and plastic, gradually replaced the traditional plaster and paper sphere.

# THE GLOBES

# 1

## TERRESTRIAL GLOBE, 1492

Martin Behaim
Painted, 51 cm (20 in.)
Germanisches Nationalmuseum, Nuremberg

Martin Behaim's globe, made in 1492, is thought to be the oldest surviving terrestrial globe. Behaim (1459–1507), a native of Nuremberg, travelled widely for commercial reasons. Trade took him to Portugal and he claimed that he had accompanied Portuguese sailors on voyages down the west coast of Africa. He returned to Nuremberg around 1490 and, conscious of the importance of these new discoveries and perhaps of the trading opportunities they might bring, he produced the globe to make them known, with the aid of local artisans. The sources for the globe are many and various. They include Ptolemaic and medieval maps, and the more up-to-date maps of the time: navigational charts known as portolan charts, which showed the latest Portuguese discoveries. Other sources of information on the globe were taken from Marco Polo (1254–1324), whose travel accounts had by then appeared in print, and Sir John Mandeville, the putative compiler of a number of travel stories written in the late fourteenth century. A map was drafted by Behaim and transferred onto the surface of the globe, in paint, by the artist George Glockenthon (d. 1514). The globe was also known as the Erdapfel (earth-apple or earth-ball). It was a colourful mixture of credible new knowledge and incredible stories from previously unknown parts of the world. Perhaps the most remarkable feature of the globe is that America is not shown, for the globe was completed before Christopher Columbus (1451–1506) returned from his westerly voyage.

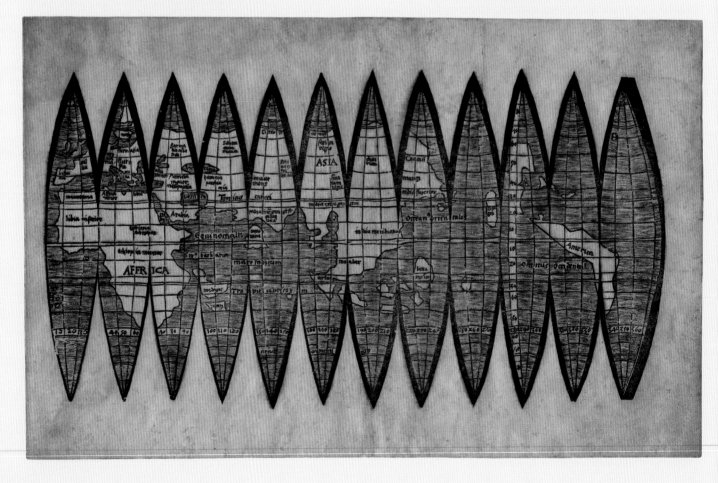

# 2

## TERRESTRIAL GLOBE GORES, 1507

Martin Waldseemüller
Woodcut
James Ford Bell Library, University of Minnesota
Facsimile globe made by Sylvia Sumira, 2007, 12 cm (4¾ in.)
Private collection

Martin Waldseemüller (1470–1521), who worked in Saint Dié in France, is probably best known for his large world map *Universalis Cosmographia* published in 1507. He also made the earliest extant printed gores, although no set mounted on a globe survives. Cartographically, both map and globe are of great significance for they show the word 'America' for the first time. Whereas Waldseemüller's map was large (248 x 136 centimetres/8 feet 1½ inches x 4 feet 5½ inches), the globe is essentially a tiny distillation of the map, with a diameter of about 12 centimetres (4¾ inches). Only a few copies of the flat gores, printed using the woodcut technique, have survived. They look crude compared with later globes, but Waldseemüller's gores show the introduction of a completely new technical and practical aspect to globe-making. To accompany the map and globe, Waldseemüller wrote a text, *Cosmographiae introductio*, which gave an introduction to cosmography, explained the principles of geometry and astronomy, and described the recent geographical discoveries illustrated on his map and, in reduced form, on his globe. The globe seen here is a modern replica made to show how the gores would have looked mounted on a sphere.

# 3

## CELESTIAL GLOBE, c. 1533–4

Johann Schöner
Woodcut, 27 cm (10½ in.)
Royal Astronomical Society, London

This celestial globe made by Johann Schöner (1477–1547) is one of only two surviving examples of the oldest extant printed celestial globe. The other example, at the Duchess Anna Amalia Library in Weimar, forms a pair with its terrestrial counterpart. The gores for this globe were printed by woodcut, and the constellations are based on those given by Ptolemy, here with their Latin names, while many individual star names are in Arabic. This globe illustrates one of the disadvantages of using the woodcut technique for fine work. It is difficult to cut small star shapes in wood, so most stars are instead shown by circular symbols. Only the brightest stars are shown with star shapes. The globe sits in a brass stand with a horizon ring and a small compass attached to one of its legs, and a plumb line ensures that the globe is level. The stand is dated 1535.

Schöner produced his first printed globes in 1515 in Bamberg, where he had started out as a cleric. In 1526 he moved to Nuremberg, where he became a renowned teacher of mathematics and again set up his own printing press, enabling him to print globe gores, as well as books and pamphlets to accompany his globes and explain their use. Research has suggested that he may have produced another globe pair in the 1520s, but no examples survive. Indeed, despite a lifetime of prolific globe production, very few examples of Schöner's globes and gores exist today.

*overleaf*
The Milky Way and several zodiacal constellations on Schöner's celestial globe. The stars and edges of the Milky Way are highlighted in gold.

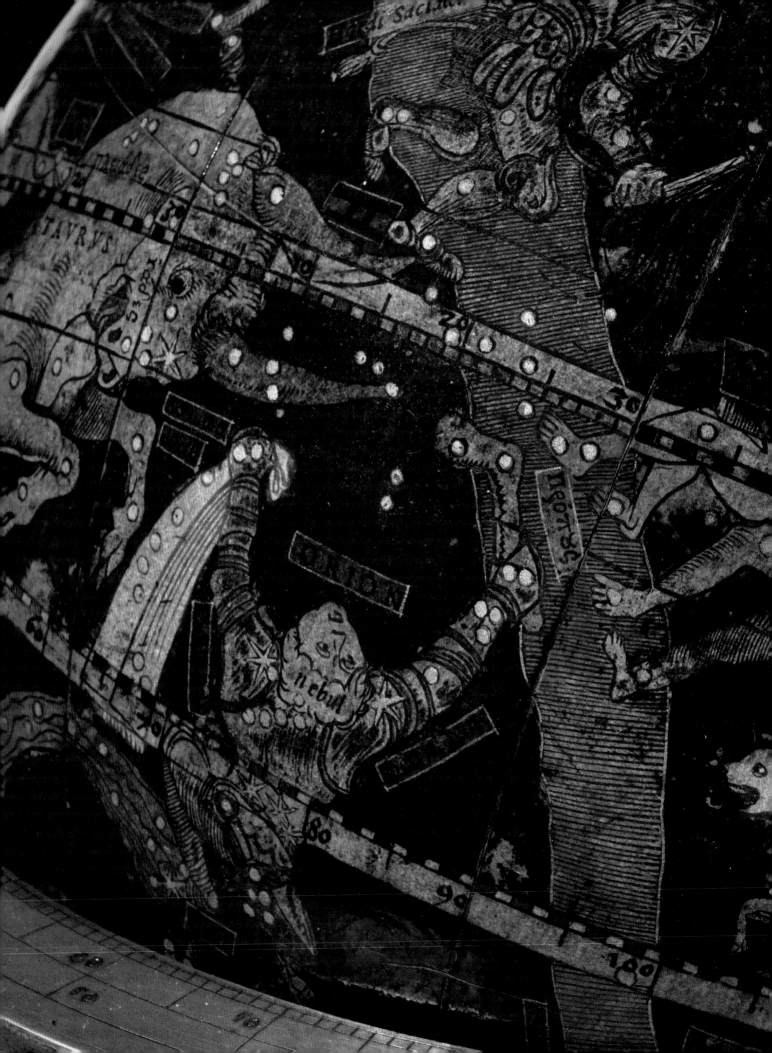

# 4

## TERRESTRIAL GLOBE, 1536

Gemma Frisius
Engraved, 37 cm (14½ in.)
Private collection, on loan to Globe Museum, Vienna

Gemma Frisius (1508–1555) was a professor of medicine and mathematics at the University of Louvain. He had a great interest in cosmography, a sixteenth-century discipline concerned with the representation of the whole visible cosmos – the heavens and the earth and the relationship between them. He made his first globe around 1530, which he described as a 'cosmographic' globe, as it showed the earth and also several important stars, but unfortunately no examples survive. He made a second terrestrial globe in 1536, followed by a celestial. The pair was made in collaboration with Gaspard van der Heyden (*c.* 1496–*c.* 1549), a goldsmith and engraver, and the young Gerard Mercator (1512–1594) who, at the time, was a student of Gemma's. All three names appear on both globes. Mercator and Van der Heyden carried out the engraving, with van der Heyden also printing the gores.

Gemma used a mixture of sources for the cartography, ranging from Ptolemy to the most recent accounts that were available from Spanish and Portuguese authors. There is a dedication on the globe to Maximilianus Transylvanus, Councillor to Emperor Charles V, from whom Gemma obtained much of the information about the Spanish discoveries. This terrestrial globe also has a selection of important stars, positioned as they would be on a celestial globe, with terrestrial longitude and latitude doubling up to give a celestial co-ordinate framework (right ascension and declination). This allowed it to be used, to a limited extent, as a celestial globe. This globe is possibly a revised version of Frisius's earlier 'cosmographic' globe, and it is the only surviving terrestrial globe by Gemma Frisius.

# 5

## CELESTIAL GLOBE, 1537

Gemma Frisius
Engraved, 37 cm (14½ in.)
National Maritime Museum, Greenwich

Gemma Frisius made a celestial partner to his terrestrial globe in 1537. The cartography of this globe was based on the woodcut maps showing the northern and southern hemispheres of the heavens made by Albrecht Dürer (1471–1528), Conrad Heinfogel (d. 1517) and Johann Stabius (c. 1460–1522) in 1515, but with several improvements. There is much more detail on the globe: for example, star brightness is indicated by six degrees of magnitude, and the names of many stars are given. In the early sixteenth century the division between astronomy and astrology was not clearly defined, and it has been suggested that this globe had an astrological use as an aid in the scheduling of medical therapies and other practices. This would have been consistent with Gemma Frisius being a professor of medicine. This globe, and its terrestrial partner, marked a significant breakthrough in globe construction. Firstly, the gores were engraved on copper plates, rather than wood, which allowed finer detail to be depicted. Secondly, copper plates were more durable than wood blocks, so more copies could be made. The plates could be also be larger, allowing the size of globes to increase. Because the stands of these globes have not survived, we do not know precisely what they looked like or how the globes were originally mounted. Gemma published no more globes after this pair, but copies of these globes were still being sold in the 1570s. This is the only surviving celestial globe by Gemma Frisius.

# 6

## TERRESTRIAL GLOBE, 1541

Gerard Mercator
Engraved, 42 cm (16½ in.)
National Maritime Museum, Greenwich

Gerard Mercator (1512–1594) was the most prolific and well-known map- and globe-maker of the second half of the sixteenth century. After he had assisted his mentor and teacher, Gemma Frisius, with a pair of globes while he was a student of the University of Louvain, he went on to make his own, the terrestrial of which appeared in 1541. At 42 centimetres (16½ inches) in diameter, it was the largest printed globe to date and the greater surface area allowed the inclusion of much more information. Although Mercator wanted to correct geographical inaccuracies that had become apparent on earlier globes, his version shows the outline of a large hypothetical southern continent. As on Gemma's globe, a selection of important stars is marked. One feature, however, makes this globe especially remarkable. For the first time, the globe carries rhumb lines or loxodromes, here emanating from thirty-two compass points. These are lines of constant bearing by which sailors could navigate to their destination. Each rhumb line cuts all meridians at the same angle, so on a globe it gently spirals towards the poles. To be able to engrave these lines on a sphere demonstrates a high level of skill. Mercator also shows the presumed location of the magnetic North Pole. Although the use of a globe in the actual practice of navigation is very limited, these elements unquestionably link the terrestrial globe with the geometry of navigation, a connection emphasised by the large pair of dividers Mercator added to the decoration of the Pacific Ocean.

*overleaf*
Dividers, rhumb lines and a star
on Mercator's terrestrial globe.

Caput Ophiuchi

260

# 7

## CELESTIAL GLOBE, 1551

Gerard Mercator
Engraved, 42 cm (16½ in.)
National Maritime Museum, Greenwich

In 1551 Gerard Mercator published what he hoped would be the most accurate celestial globe of its time. Recent studies have shown that he used the Copernican theory of the heliocentric universe to find the correct position of the stars for the epoch 1550. Again, he improved on the work of his predecessors and added constellation figures for the asterisms (groups of stars smaller than a constellation) Antinous and Cincinnus (Coma Berenices). Much of the area around the South Pole is blank, as it had yet to be charted. Latin and Greek are used to name the constellations.

In a change from earlier globes, Mercator designed his celestial gores to align with the equatorial co-ordinates rather than the ecliptic co-ordinates, so that the gores met at the celestial poles, which coincided with the support axis of the sphere. As with his terrestrial globe, he clipped the gores at 70 degrees latitude and covered each polar area with a circular cap or calotte. The size of this globe (like its terrestrial partner) made it impractical to mount it in the metal tripod-type stand that supported some earlier globes. Mercator's design of a wooden stand with a circular base, with four legs supporting the horizon ring, provided an elegant, practical and adaptable solution for the mounting of globes, and came to be used repeatedly for the next 400 years.

Antinous

280

290

300

SAGITTARIVS

Βοώτης
φύλαξ

Cincinnus,
Cæfaries; πλόκαμος,
Berenicis crinis, Trica
☽ ♀

Tropicus Cancri

VIRGO παρθενος
Erigone
Preuindemiator προ
μήτηρ Al

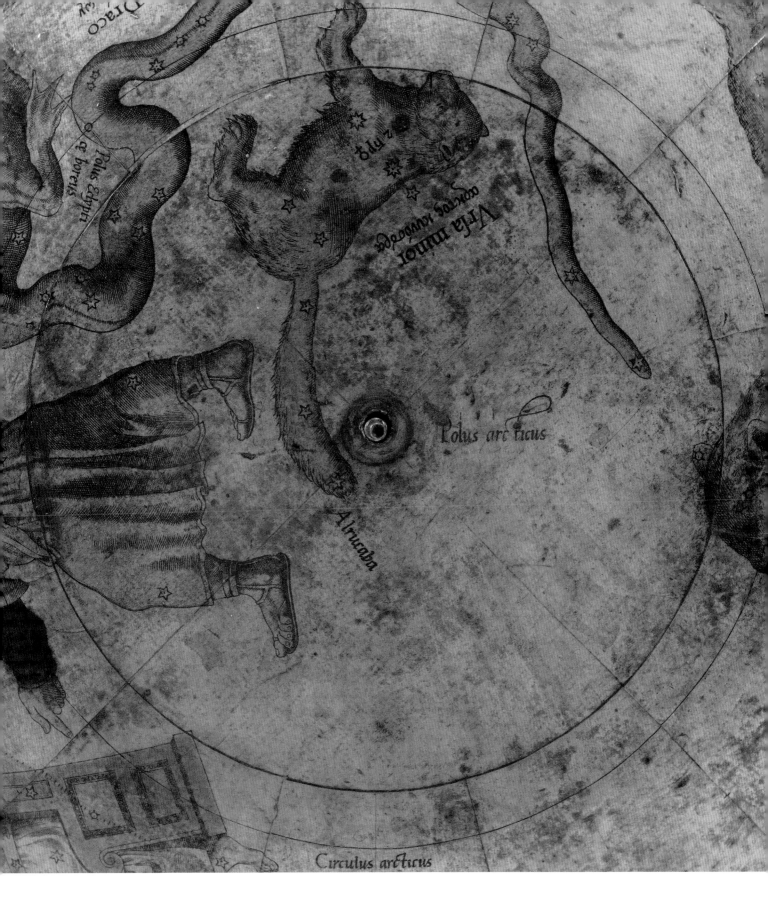

Draco

Polus celypti e boreus

Polus Egypti

Vrſa minor

ocetros κυνοσȣρα

Polus arcticus

Alrucaba

Circulus arcticus

*opposite*
Two asterisms – Antinous (top) and Cincinnus
(bottom) – introduced by Mercator.

*above*
The celestial north pole of Mercator's celestial globe,
with its circular paper cap.

THE GLOBES    61

# 8 AND 9

## TERRESTRIAL AND CELESTIAL GLOBE GORES, c. 1560

François Demongenet
Engraved, 8.5 cm (3¼ in.)
British Library, London

These early gores were made by François Demongenet (active 1550–60), a contemporary of Gerard Mercator, who lived in the town of Vesoul in eastern France. He was a mathematician and map-maker and, in 1552, he published a set of woodcut terrestrial and celestial gores that could be used to make small globes with a diameter of about 8.5 centimetres (3¼ inches). Several years later, around 1560, he published a set of revised gores of the same size, but this time they were engraved on copper plates.

As they were so small, not a great deal of information could be shown on these gores, but some of the inscriptions and imagery suggest that Demongenet had copied the globe gores of 1547 by the instrument-maker Georg Hartmann (1489–1564), who in turn had been influenced by Gemma Frisius's globes. Demongenet is not very well known as a globe-maker, but his gores of 1560 are significant, because they are the source for many of the extremely fine silver and gilt globes made in the latter part of the sixteenth century. These highly decorative globes were engraved by hand directly onto the metal surface. They were of a similar size to Demongenet's gores, which would therefore have been relatively easy to copy. Among the craftsmen making these globes were Georg Roll (1546–1592) and Johann Reinhold (c. 1550–1596), who became well known for their elaborate clockwork pieces incorporating globes. These were expensive, luxury items made as showpieces for the very rich (see Globe 15, p. 86).

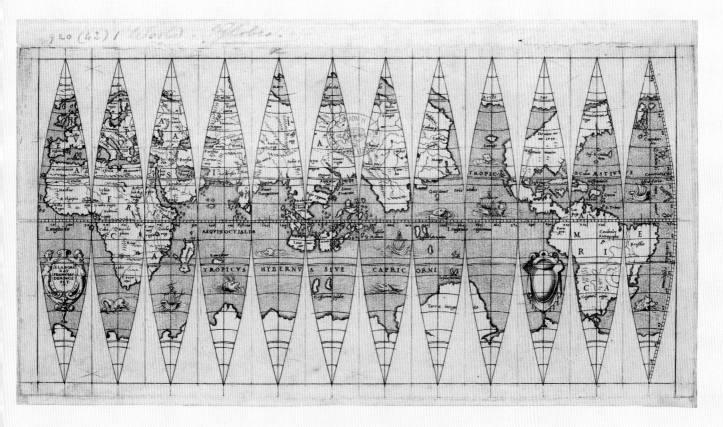

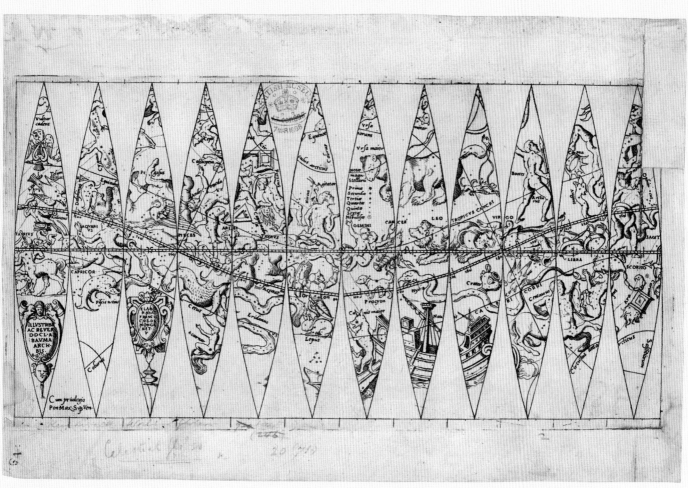

# 10

## TERRESTRIAL GLOBE, 1592/1603

Emery Molyneux
Engraved, 63.5 cm (25 in.)
Honourable Society of the Middle Temple, London

Emery Molyneux (d. 1598) produced the first printed globes in England. At 63.5 centimetres (25 inches) in diameter, they were the largest printed globes to have been made anywhere at the time, and were clearly designed to make an impact. The terrestrial globe, first published in 1592, highlighted the glories of recent English maritime enterprise and achievement. The tracks of the circumnavigations of Sir Francis Drake and Thomas Cavendish are shown, and the discoveries of John Davis (*c.* 1550–1605) and Sir Martin Frobisher (*c.* 1535–1594) in the North Atlantic, and Sir Walter Raleigh (*c.* 1554–1618) in Virginia, are also depicted. The globe is dedicated to Elizabeth I, and the wording suggests that the globe would enable her to see at a glance how much of the seas her naval forces could control. In this way, the globe was a potent symbol of power.

The making of these magnificent globes was a collaborative effort. They were financed by a rich merchant, William Sanderson (*c.* 1548–1638), who had funded and indemnified many of the voyages and discoveries shown on the globe. Molyneux, who had some seafaring experience himself, gathered together the most recent maps, charts and accounts of voyages. He employed Edward Wright, a mathematician, to plot the coastlines, and Jodocus Hondius the Elder (1563–1612), a highly skilled Dutch émigré, to carry out the engraving. The terrestrial globe at the Middle Temple is dated 1603 and differs slightly from the only other surviving example, which bears the date 1592. Several revisions have been made to the geography in the areas of the Arctic and the west coast of America. Perhaps Hondius made the revisions, but we do not know for certain.

*overleaf*
A note in the Atlantic Ocean states that Virginia was first inhabited and farmed by the English. Sir Walter Raleigh, who was instrumental in the colonisation of North America, is named.

*on pages 68–69*
Sir Francis Drake's track. The track above, picked out in green, is that of Thomas Cavendish. A sea monster and a man-like sea creature are also shown.

*Virginia primum lustrata, habitata &*
*culta ab Anglis. Impensis D. Gualteri de*
*Ir Ralegh Equitis Aurati &c. Annuente*
ELIZABETHA S.r Angliæ Regina.

Montagnas
Henrico
iglas

Bermuda

O C E A

Heptopolis

A T L A

Anegade

Lonbiero
S. Bertolomeo
Martin
S. Bernardo
Saba

Luggi

TA...A...

Mogar

Benama

taxa

Zimbaos

Dofa

de S. Laurenzo

C. de S

Ionros

Dauagul

Rio de lago primum
nunc R del Spirito S.a.

Samot

Valuta

Belugaras

R. d los Reyes

REG.

Punta de S. Maria

S. Lucia

Vigiti magna

Alagoa

ꝟ os Medanos

Corrad

Tiscara

R. de infante

Punta primira de nauidad

Il. Chaons

Penna

das fontes

MA HA

M

B. dalagoa

BA NA

C. de Avecife

Reg.

R. de lagnna

Bara hermofa y

Pefqueria

C. de las vacas

Infantes

S. F. Drack

# 11

## CELESTIAL GLOBE, 1592/1603

Emery Molyneux
Engraved, 63.5 cm (25 in.)
Middle Temple Library, London

Like its terrestrial partner, this celestial globe bears a dedication to Elizabeth I. It was remarked at the time that it did not differ greatly from the celestial globe by Gerard Mercator, with the exception of the addition of two new constellations in the southern hemisphere, the Southern Cross and the Southern Triangle. These constellations first appeared on a new globe by Jacob Floris van Langren (*c.* 1525–1610) and Petrus Plancius (1552–1622), published in Amsterdam in 1586, which Molyneux essentially copied, and which had in turn been based on Mercator's globe of 1551. The positions of the new constellations, however, were too far west. Jodocus Hondius the Elder, the engraver of Molyneux's globes, corrected the positions when he made his own globe in 1600.

Many contemporary accounts show that the globes were well known at the time. Thomas Bodley (1545–1613) acquired a pair for his new library in Oxford in 1602, and the Warden of All Souls College, Oxford, also bought a pair. It seems they were made in large numbers and could cost up to £20 a pair, equivalent to at least several thousand pounds today. As described in a treatise written by Robert Hues to accompany the globes, first published in 1594, a smaller version was also made; 'so are they of a cheaper price: that so the meaner Students might herein also bee provided for.' No examples of the smaller globe exist, and very few examples of the larger globes survive. In addition to the pair at the Middle Temple and the terrestrial example dated 1592 at Petworth House, West Sussex, there are only two celestial globes in Germany.

*overleaf*
The dedication to Elizabeth I
and the new constellation 'Crux',
the Southern Cross.

Deneb kaitos

DIVÆ
ELISABETHÆ
Potentissimæ & Sere-
nissimæ Angliæ, Franciæ &
Hiberniæ Reginæ, fidei defensatrici Guilielmus
Sanderson globum hunc cœlestem, suis etiam sumpti-
bus, operaq́ Emery Molineux, feliciter
elaboratum animi eius obseruantia antea
et deuotissuis mer
B. B. D.

Judocus Hon-
dius Fen. Sc.
1592

Acarnar

Beรนน

CRVX

Elcapo

# 12

## TERRESTRIAL GLOBE, 1606/21

Willem Jansz Blaeu
Engraved, 13.5 cm (5¼ in.)
British Library, London

Willem Jansz Blaeu (1571–1638) was one of the leading globe- and map-makers of the seventeenth century. As a young man he had an interest in astronomy and spent several months studying with Tycho Brahe (1546–1601), the renowned Danish astronomer. Blaeu settled in Amsterdam in 1598 and established a publishing firm, which, in addition to globes, produced maps, books and atlases. His first terrestrial globe was published in 1599 and had a diameter of 34 centimetres (13¼ inches). A celestial partner followed shortly after, and he went on to make globes with diameters of 10, 13.5 and 23 centimetres (4, 5¼ and 9 inches), culminating with a large pair, 68 centimetres (26¾ inches) in diameter, first published in 1617. Blaeu's globes were repeatedly updated as new information came to light.

This globe was originally published in 1606 but has been revised to include the discoveries at the tip of South America in 1616 by the Dutchmen Willem Schouten (active 1590–1618) and Jacob Le Maire (1585–1616). The only known route at the time from the Atlantic to the Pacific Ocean was through the Magellan Strait, which was monopolised by the Dutch East India Company, so the 1616 expedition set out to find an alternative route. They succeeded by discovering a strait between Tierra del Fuego (which they established was an island) and another southerly island. They named the strait after Le Maire, and the new island, Staten Lant (Staten Island), after the States General, which had licensed their voyage. The pair went on to discover Cape Horn, which they named after Hoorn, the Dutch town from where most of the finance for the expedition came.

*overleaf*
The Strait of Le Maire,
Staten Island and Cape Horn.

# 13

## CELESTIAL GLOBE, 1606/21

Willem Jansz Blaeu
Engraved, 13.5 cm (5¼ in.)
British Library, London

Blaeu's first globe was celestial. He made it around 1597–8 and used Tycho Brahe's observations of the fixed stars as its cartographic base. The style of his celestial globe was different from those of his contemporaries and predecessors, for he did not engrave the globe himself but employed Jan Pietersz Saenredam (1565–1607), who has been credited with creating this new style, which was much more decorative and introduced changes in the depiction of figures. For example, in the northern constellation Bootes, the figure is dressed for a cooler climate with a coat, fur-trimmed hat and boots, rather than a more classical costume. These innovations were copied in new editions of globes by Blaeu's competitors. Competition with his rival Jodocus Hondius was fierce in the early 1600s, and they both produced several new editions of globes in rapid succession, each claiming superiority over the other. The globe pictured here is a smaller version of Blaeu's earlier globes. A note in the cartouche tells us that the positions of the fixed stars were adjusted to the year 1606 (using the observations of Tycho Brahe), and that the globe also shows the stars in the southern hemisphere that had been observed by Frederik de Houtman (1571–1627), who had recently returned from a voyage to the southern seas. Also included on the globe is a nova (a star showing a rapid increase in brightness, sometimes thought to be a new star as it becomes visible) in the constellation Cygnus, which Blaeu had personally observed in 1600.

*overleaf*
The new star in the constellation Cygnus.

# 14

## THE CHINESE GLOBE (TERRESTRIAL GLOBE), 1623

Painted, 59 cm (23 in.)
British Library, London

Called 'the Chinese Globe', and the earliest extant terrestrial globe known to have been made in China, this unique and rather beautiful terrestrial globe differs from the other globes in this book because it is painted rather than printed, and was not made in the West, but it shows a coming together of Western Renaissance thought and Chinese culture. Jesuit missionaries had been active in China since the late sixteenth century and, in addition to religion, they brought with them a European understanding of science. They were, however, eager to learn about Chinese culture and engage with scholars at the Ming Court. Manuel Dias (1574–1659) and Nicolo Longobardi (1565–1655), to whom the globe is attributed, were Jesuit scholars who became highly respected at court. Their names appear on the globe as Yang Ma-no and Long Hua-min. It is not clear whether the globe was commissioned by the Emperor or whether it was made as a gift. The globe gives a Western view of the world as it was then known, which was quite different from Chinese maps that showed China at the centre but gave little information about the rest of the world. The text is in Chinese and it explains the concepts of longitude and latitude, and how astronomical observations can be used. It also incorporates references to Chinese ideas about magnetism, which preceded Western understanding of this phenomenon.

There are references to Chinese celestial globes dating back to the fifth century AD, but none has survived. There is also an account of an earlier terrestrial globe, which has not survived, having been made in 1603, again under the guidance of the Jesuits.

# 15

## CLOCKWORK CELESTIAL GLOBE, 1646

Isaac Habrecht III
Engraved on copper-gilt, 19 cm (7½ in.)
National Maritime Museum, Greenwich

During the second half of the sixteenth and the first half of the seventeenth century, a number of finely engraved and sophisticated clockwork-driven metal celestial globes were made. These globes could be constructed only because of recent developments in horology, in particular the mainspring and fusee for regulating its force, which enabled clocks to be smaller and portable. They were not primarily timepieces; rather, the mechanism enabled the sphere to demonstrate the movement of the heavens. Though it has been postulated that these globes were used for making astrological predictions, it is much more likely that their function was more decorative: that they were regarded as extremely expensive showpieces of current technology and as status symbols for those who could afford such items.

Isaac Habrecht III (1611–1686) belonged to a family of clock- and globe-makers in Strasbourg. His grandfather Isaac Habrecht I (1544–1620) helped build the astronomical clock in Strasbourg Cathedral and his uncle Isaac Habrecht II (1589–1633) published a pair of printed globes in the early 1620s. Whereas many earlier clockwork globes were based on the gores of Mercator or Demongenet, the design of the constellations on Habrecht's globe seems to be based on the celestial globe of Johannes Oterschaden, who worked in Strasbourg around 1600.

Clockwork mechanisms were applied to terrestrial globes at a later stage, but in the early sixteenth century it was still generally thought that the earth was static, and hence there was little point in creating a rotating terrestrial globe.

# 16

## POCKET GLOBE, c. 1679

Joseph Moxon
Engraved, 7 cm (2¾ in.)
British Library, London

After the success of the globes made by Emery Molyneux at the end of the sixteenth century, no new globes were produced in England until Joseph Moxon (1627–1691) revived the art in the 1650s. He published his first globes in 1653, complementing them with manuals on their use. Both his globes and his books were very popular. There were five editions of his manual of the use of globes, the last being published in 1698, after Moxon's death, by his son, James. The globes were made in several sizes, but it is for one of his smallest globes that Moxon is best known. This is the 'pocket globe', which first appeared in the 1670s. In a catalogue of items advertised for sale, Moxon describes the arrangement as 'Concave hemispheres of the starry orb, which serves for a case to a terrestrial globe of 3″ diameter made portable for the pocket.'

The terrestrial pocket globe shows the tracks of Sir Francis Drake and Thomas Cavendish, and California is depicted as an island. The celestial gores are concave, but in fact the constellations appear reversed from the way we see them when we look up from the earth, as though they have been drawn to fit a common celestial globe. These globes were attractive and very popular, and as much an amusing gadget as a geographical tool; and, priced at 15 shillings, they were much cheaper than a pair of globes. The language of this globe is Latin, but Moxon also made an English version. His pocket globes started a tradition in England, and subsequent globe-makers continued to produce them until well into the nineteenth century. Moxon's globes are now very rare.

# 17

## TERRESTRIAL GLOBE, 1688/1752

Vincenzo Coronelli
Engraved, 108 cm (42½ in.)
National Maritime Museum, Greenwich

The Franciscan monk Vincenzo Coronelli (1650–1718) is one of the greatest of all globe-makers. In 1681 he was commissioned to make a large pair of painted globes for King Louis XIV of France. While in Paris, preparing to make the globes, Coronelli had access to a vast amount of geographical and astronomical information, which he was able to use later in his own work. He also built up a network of geographers, travellers and scholars throughout Europe, who were able to provide up-to-date source material for him. After this undertaking he returned to his home town of Venice and set up a workshop to make globes, maps and atlases in the convent of Santa Maria Gloriosa dei Frari. He had the backing of the Venetian Senate and was granted the title 'Cosmographer of the Serene Republic'. One of his many projects was to publish a reduced, printed version of the French king's globes. At 108 centimetres (42½ inches) in diameter, they were the largest printed globes at the time and for many years to come, and they enabled the inclusion of an impressive amount of geographical and historical detail. The production of the globes was funded by subscribers to the society he had founded in 1684, the Accademia Cosmografica degli Argonauti, which was the first geographical society in the world. Its members were wealthy people who wanted to have, and could afford, resplendent globes. The first edition of the globe was printed and published in Venice in 1688, but many of the gores for these large globes were reprinted at a later date, and they were sometimes mounted onto spheres even later on. A note inside this particular globe states that it was made in Vienna by a Franciscan monk, Tobias Eder, with help from Matthias Heinen in 1752.

# 18

## CELESTIAL GLOBE, 1693

Vincenzo Coronelli
Engraved, 108 cm (42½ in.)
British Library, London

The history of Vincenzo Coronelli's large celestial globes is complicated, as several editions were published and printed in both Paris and Venice. After the success of the large painted globes he had made for Louis XIV, Coronelli was granted a privilege by the king in 1686 to engrave and print maps in Paris for a period of fifteen years. Coronelli had only just founded his workshop in Venice, and it lacked sufficient skilful workers to be able to fulfill his ambitious plans for the engraving of both his large globes. Therefore, using his privilege, he sought help in Paris and negotiated a contract with Jean Baptist Nolin (1657–1725), engraver to the king, to work on the large celestial globe, for which Coronelli supplied the draft maps. This contract enabled the first edition of the celestial globe to appear at the same time as the first edition of the terrestrial globe in 1688. In 1693 Nolin published a new edition, which he dedicated to Coronelli. At the same time in Venice, the skills of Coronelli's workforce had greatly improved and he issued a Venetian edition of the celestial globe (and also a new edition of the terrestrial globe). The globe illustrated here is an example of the 1693 Paris edition.

# 19

## *LIBRO DEI GLOBI, 1697/1701*

Vincenzo Coronelli
British Library, London

Vincenzo Coronelli was keen to promote and sell his globes throughout Europe and to show their superiority over those of his predecessors. Through the geographical society he founded he raised advance funds by subscription for the making of his globes and, by 1697, he was making globes in five sizes ranging from 5 to 108 centimetres (2 to 42½ inches). However, globes are not as portable as maps and atlases, and it was therefore difficult for other potential clients and anyone interested in the globes to see them and assess their quality. To reach a wider audience, Coronelli had the idea of collecting together the gores of his full range of globes and binding them all in one volume, which was essentially a catalogue. In this way he could reach a wider audience. The *Libro dei Globi* was published in 1697 and formed part of his grand project, the *Atlante Veneto*, which consisted of a series of atlases and was published in instalments from 1691 onwards. In addition to being a record of Coronelli's full range of globes, including references to the grand globes made for Louis XIV, the book was a clever way of advertising his work. Furthermore, it was cheaper to buy the book than a pair of large globes. The few copies of the *Libro dei Globi* that survive are all slightly different from each other in the number or arrangement of the plates. The copy at the British Library appears to be an early example of the second edition of 1701.

# GLOBI
*Differenti*
*Del*
# P. CORONELLI,
*Consacrati*
A' SUA ECCELLENZA
*Il Signore Conte*
*Di*
# LAMBERG
*Ambasciatore Cesareo,*
*et c.*
MDCCI.

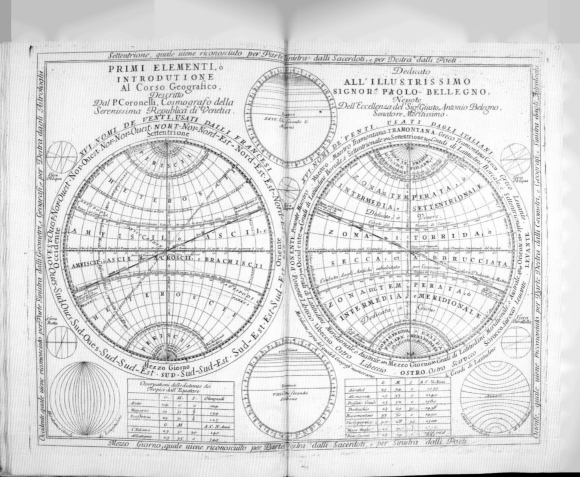

**PRIMI ELEMENTI, ò**
**INTRODUTIONE**
Al Corso Geografico,
*Descritto*
Dal P. Coronelli, Cosmografo della
Serenissima Republica di Venetia.

*Dedicato*
**ALL'ILLUSTRISSIMO**
**SIGNOR. PAOLO. BELLEGNO,**
*Nepote*
Dell'Eccellenza del Sig. Giusto Antonio Belegno,
Senatore Meritissimo.

*Mezzo Giorno, quale viene riconosciuto per Parte Destra dalli Sacerdoti, e per Sinistra dalli Poeti.*

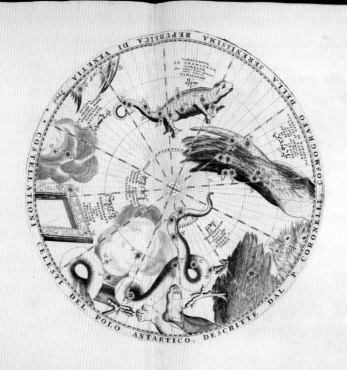

COSTELLATIONI CELESTI DEL POLO ANTARTICO, DESCRITTE DAL P. CORONELLI

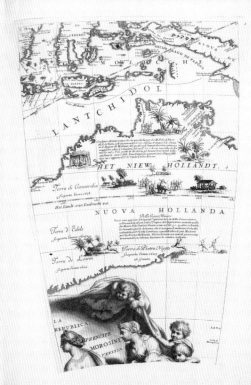

*opposite above*
Pages from the *Libro dei Globi*.

*opposite below*
South polar cap for the 108 cm
(42 in.) celestial globe of Coronelli.

*above*
Terrestrial gore sections for the
108 cm (42 in.) terrestrial globe
of Coronelli.

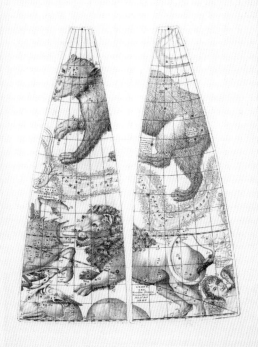

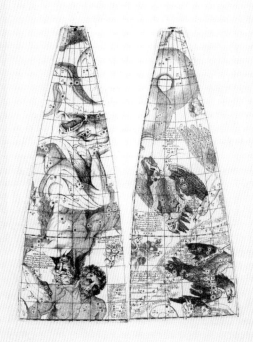

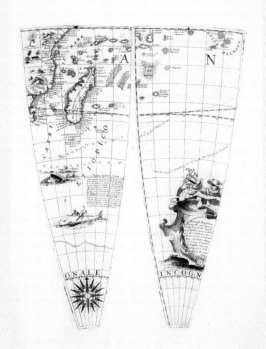

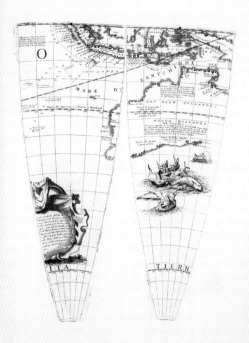

*above and opposite*
Gore sections for the 47 cm (18½ in.) globes of Coronelli.

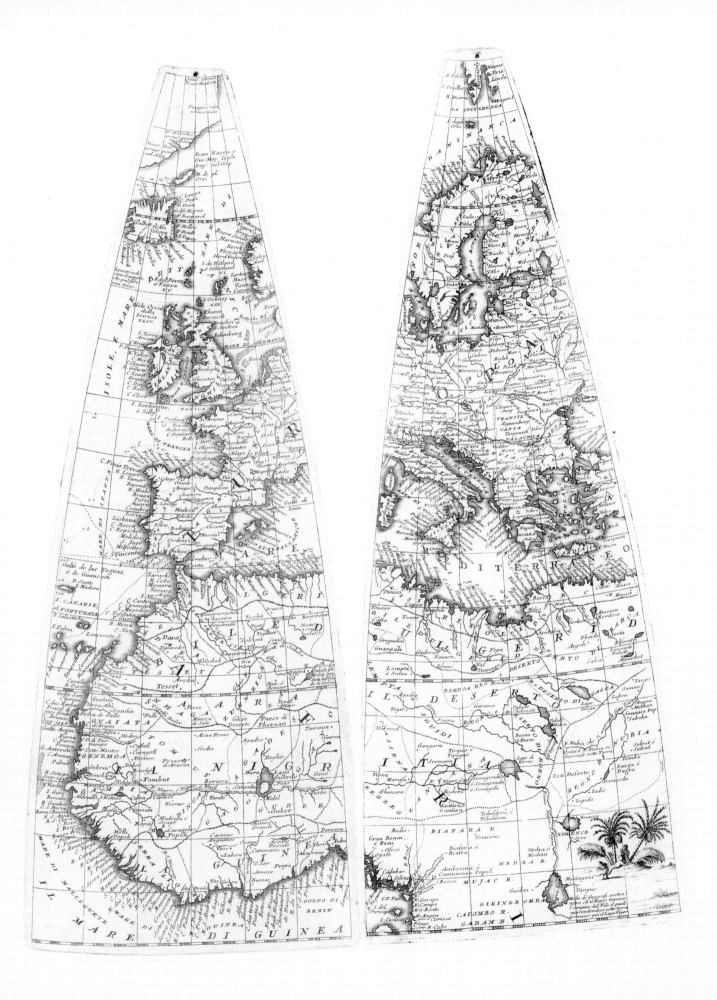

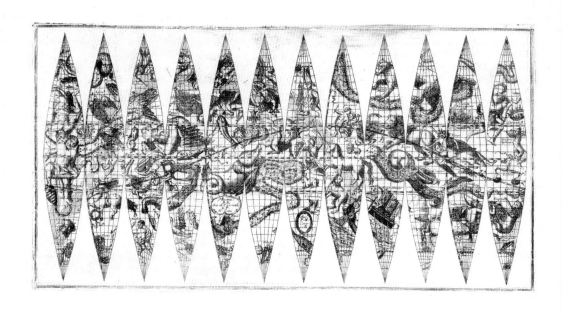

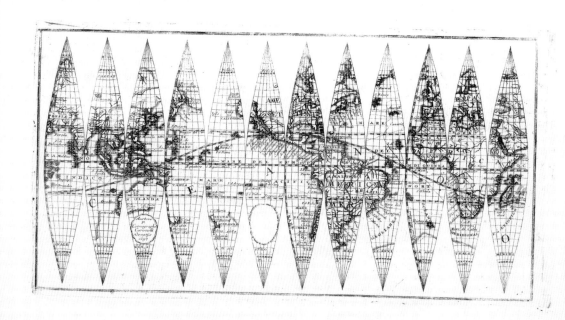

Gores for the 8.5 cm (3½ in.) globes of Coronelli.

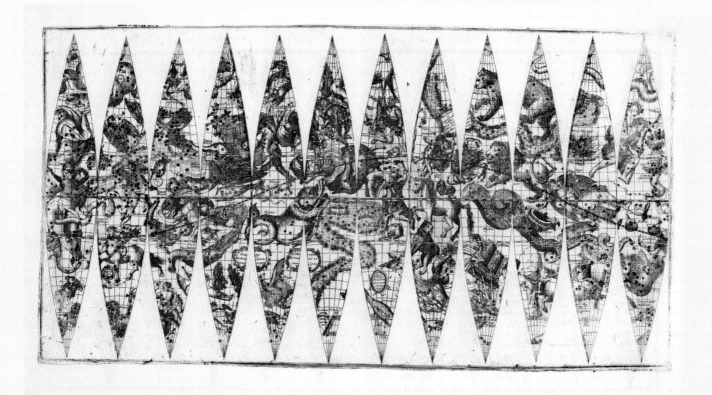

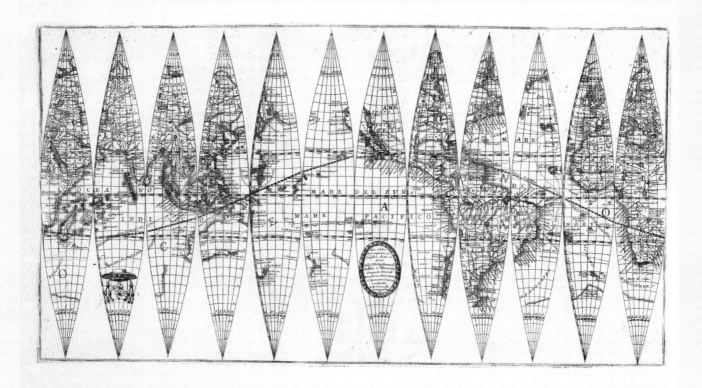

Gores for the 15 cm (6 in.) globes of Coronelli.

# 20

## CELESTIAL GLOBE, 1700

Thomas Tuttell
Engraved, 36 cm (14 in.)
British Library, London

Thomas Tuttell (*c.* 1674–1702) was a highly regarded instrument-maker and was appointed hydrographer (responsible for sea charts) and mathematical instrument-maker to the king, William III. At his workshop at the sign of the King's Arms and Globe at Charing Cross, he sold a variety of instruments, including globes. Though his name appears on a label pasted onto this globe, he almost certainly did not make it. It is identical to a celestial globe made by Joseph Moxon in 1653, except that this one has an additional constellation in the northern hemisphere, Cor Caroli (Heart of Charles), which refers to King Charles I (see pages 40–1). It was named by Sir Charles Scarborough to commemorate the restoration of the monarchy in 1660 and it was first published on a star map by Francis Lamb in 1673. Tuttell is known to have worked with James Moxon, son of Joseph Moxon, on a book entitled *Mathematicks made Easie* around 1700, and this connection makes it likely that he would have been able to acquire globes by Moxon. It was not uncommon for globe- and instrument-sellers to paste their own labels onto instruments made by other people. Removing the label might reveal Joseph Moxon as the maker.

# 21

## POCKET GLOBE, 1710

Herman Moll
Engraved, 7 cm (2¾ in.)
British Library, London

Herman Moll (1654–1732) – a map-maker and engraver, probably of German descent, who worked in London from 1678 – published only one globe, a pocket globe. The prime meridian goes through Ferro in the Atlantic Ocean, California is shown as an island, and the track of William Dampier's (1651–1715) first circumnavigation, between 1679 and 1691, is depicted in red. Dampier was a buccaneer but also a keen observer of people and nature on his travels. He was the first Englishman to circumnavigate the world since Thomas Cavendish a hundred years previously, and was also the first person to do this three times. In 1688, he was on the first English ship to touch the coast of New Holland (Australia).

After Dampier returned to England he published an account of his travels, *A New Voyage Around the World* (1697), which became a bestseller. The cartouche on his pocket globe has the inscription 'A Correct globe with ye Trade-winds by H. Moll'. Dampier had taken an interest in winds and it is probable that the depiction of trade winds was an innovation inspired by his observations. The cartouche on the celestial map on the inside of the case says 'A Correct Globe With ye New constellations of Mr. Hevelius 1710'. Johannes Hevelius's (1611–1687) new star catalogue was published posthumously by his wife, Elisabeth, in 1690. The Polish astronomer had introduced eleven new constellations, which are depicted on Moll's globe.

# 22

## POCKET GLOBE, c. 1715

Johann Baptist Homann
Engraved, 7 cm (2¾ in.)
British Library, London

Shortly after 1700 the map-maker Johann Baptist Homann (1664–1724) founded his own publishing house in Nuremberg; it went on to become the most important publisher of maps and atlases in Germany. After his death his son carried on the business, which lasted well into the nineteenth century. Homann collaborated with several globe-makers, among them Johann Gabriel Doppelmayr (1671–1750) and Christoph Eimmart (1638–1705). Homann published only one pair of globes himself, in the form of a pocket globe. As is usual, the terrestrial globe is housed in a leather case and lined with celestial gores, which are concave, so we see the heavenly constellations as from the earth. However, with this example, rather than being solid, the terrestrial globe is a hollow wooden sphere. The northern and southern hemispheres can be pulled apart and inside the globe we find a miniature armillary sphere made from card, with an illustrated zodiacal band, complete with a sun at the centre. The legend in the cartouche informs us that recent observations by the Académie Royale des Sciences in Paris were used in the plotting of both globes. On the terrestrial globe, as we have seen elsewhere, California is shown as an island.

# 23

## TERRESTRIAL GLOBE, 1728

Johann Gabriel Doppelmayr
Engraved, 32 cm (12½ in.)
British Library, London

Johann Gabriel Doppelmayr (*c.* 1677–1750), professor of mathematics in Nuremberg, had a wide-ranging interest in the Enlightenment concerns of natural philosophy (an eighteenth-century term for science), including electricity. He travelled extensively, spoke several languages, and was a member of several scientific societies, including the Royal Society in 1733. He was a friend of Johann Baptist Homann, and they collaborated on several atlases and books.

Doppelmayr's first pair of globes appeared in 1728; they were followed by a second pair of 20 centimetres (7¾ inches) in 1730 and a pair of 10 centimetres (4 inches) in 1736. His globes were engraved by Johann Georg Puschner (1680–1749), who was also known as an instrument-maker. The terrestrial globe displays many of the recent discoveries of the early eighteenth century, but it also pays homage to several important explorers of the past. Many tracks are shown, including those of Magellan, Le Maire, Tasman and Dampier, as well as the very recent journey of the Dutchman Jacob Roggeveen (1659–1729), who discovered Easter Island in 1722 during an unsuccessful expedition to find the mythical great southern continent. An address to the reader is surrounded by twelve charming portraits of famous explorers with Martin Behaim, a fellow citizen of Nuremberg and globe-maker, in the most prominent position. Doppelmayr had re-established globe-making in Nuremberg.

*overleaf*
The address to the reader
surrounded by portraits
of eminent explorers.

<h1 style="text-align:center">24</h1>

## CELESTIAL GLOBE, 1728

Johann Gabriel Doppelmayr

Engraved, 32 cm (12½ in.)

British Library, London

Long before he published his first celestial globe in 1728 Doppelmayr had taken a keen interest in astronomy, and he spent some time studying the subject in Leiden, one of the leading universities of the time. After his studies he returned to his native town of Nuremberg, and, as a teacher, was very active in promoting new scientific ideas. In the early 1700s he had compiled several celestial maps, which had been published in various atlases by his friend Johann Baptist Homann. These maps were later collected and published in 1742 as the *Atlas Novus Coelestis*, for which Doppelmayr became well known. He also translated several scientific works into German, including Nicolas Bion's *L'usage des globes célestes et terrestres* (1699) and John Wilkins's *Discovery of a World in the Moone* (1638), which advanced the relatively new theories of Copernicus and Galileo. Doppelmayr's celestial globe was accurate for the epoch 1730 and, like Herman Moll's, it drew on the star catalogue of Johannes Hevelius of 1690. Also depicted are the paths of several comets observed by Hevelius, Johann Kepler, Giovanni Cassini and John Flamsteed. There were other German globe-makers in the early 1700s but Doppelmayr's globes dominated the German market until the end of the eighteenth century. They were revised in the 1750s and finally in 1792 by Wolfgang Paul Jenig (d. 1805), forty-two years after Doppelmayr's death.

*overleaf*
The paths of comets on
Doppelmayr's celestial globe.

# 25

## POCKET GLOBE, c. 1730

John Senex
Engraved, 7 cm (2¾ in.)
British Library, London

John Senex (1678–1740) was a publisher of scientific tracts and a prolific map- and globe-maker. He dominated the globe trade in England in the first part of the eighteenth century. In the early part of his career he collaborated with several other globe-makers, but around 1710 he moved to premises in Fleet Street, London, and the globes he made from then on were in his name only. He continued to work with others involved in the globe trade, for example the Cushee family, but only on maps and surveys. His globes ranged in size from 7 to 68 centimetres (2¾ to 27 inches). He was elected a Fellow of the Royal Society in 1728 and his globes after this date prominently and proudly bear the initials FRS.

For the celestial gores lining the case of the example illustrated here, Senex reused the gores he had made with Charles Price (*c.* 1679–1733) around 1710. For the terrestrial globe, the gores were freshly engraved and the style of the lettering has changed, but they differ only slightly from those Senex made with Price. As with the globe of 1710, the prime meridian goes through London, and California is depicted as an island. The trade winds, however, are shown with greater clarity. The Antipodes of London – the place directly opposite to London in the southern hemisphere – are pinpointed with a circle of letters and Drake's track has been removed. Globe-makers would continue to use the copper plates for this globe long after Senex died. The plates were revised to show California as a peninsula and George Anson's circumnavigation of 1744, and they were bought by another globe-maker, George Adams (1709–1772), in the 1750s.

# 26

## POCKET GLOBE, 1731

Richard Cushee
Engraved, 7 cm (2¾ in.)
British Library, London

Richard Cushee (1696–1733), a map engraver and land surveyor, may be one of the lesser-known globe-makers of the early eighteenth century, but he had connections with many others in this trade. An apprentice of Charles Price, who had collaborated with John Senex, Cushee in turn took on Nathaniel Hill (active 1742–68) as his apprentice and they all made globes. Cushee also collaborated with Thomas Wright (1692–1767), an instrument-maker, and together they were involved in the publication of a book on the use of globes, *The Description and Use of the Globes and the Orrery,* by the astronomer and assayer to the Royal Mint, Joseph Harris (*c.* 1704–1764). The first edition was published in 1731 and it proved to be very popular; there were seven reprints during Cushee's lifetime and it continued to be printed after his death. By 1793 had been reprinted fourteen times.

In the same year as the first edition, Cushee published a pocket globe, for which he is probably best known. As usual, he housed the terrestrial globe in a case, with celestial gores pasted on the inside. And yet, in a departure from most previous pocket globes, Cushee lined his case with concave gores that showed the constellations as we see them from the earth. This contrasts with the pocket globe of John Senex (Globe 25) published only one year earlier. The difference is most obviously seen in the constellation Ursa Major (the great bear), where his snout points in different directions on each globe.

The concave celestial gores of Cushee's pocket globe were later used again by Nicolas Lane, who acquired the copper plates in the 1770s (Globe 35, p. 155).

# 27

## TERRESTRIAL GLOBE, 1730

Richard Cushee
Engraved, 30.5 cm (12 in.)
British Library, London

Richard Cushee also made a pair of globes 30.5 centimetres (12 inches) in diameter, dated 1730, and an undated pair 38 centimetres (15 inches) in diameter. There is an advertisement for them in Joseph Harris's book on the use of globes (Figure 17). The globes are boldly engraved and incorporate much recent detail. Monsoon winds are shown, and there are new North American additions, with the East Coast being described in greater detail. In contrast, below the North Pole it clearly states 'Parts Unknown'. The region around New Holland (Australia) is based on Dutch maps. Surprisingly, in the Pacific Ocean to the west of Mexico a tiny island, 'Cap. Clippertons I.', is prominent (see overleaf). It was named after John Clipperton (d. 1722), a pirate who had led a mutiny against William Dampier in a privateering voyage of 1703. There are many little pictures of ships scattered throughout the oceans, though no tracks are shown. The prime meridian passes through London. After the founding of the Royal Observatory at Greenwich in 1675, astronomers and navigators began to use Greenwich as the starting point for the measurement of longitude, and this comes to be reflected in maps and globes.

# 29

## TERRESTRIAL GLOBE, 1757/1770

John Senex and Benjamin Martin
Engraved, 30.5 cm (12 in.)
British Library, London

After the death of John Senex in 1740, his widow, Mary, continued the business successfully until 1755, when she decided to retire and sold the stock. Apart from the apparatus needed to make the smallest globe, all the globe-making equipment was bought by James Ferguson (1710–1776), an enterprising self-taught Scotsman who came to London in 1743 and became well known for his lectures on scientific matters, as well as for his orreries and other models. Overwhelmed by globe-making in addition to his other activities, Ferguson sold his plates in 1757 to Benjamin Martin (1705–1782), an instrument-maker who also gave public lectures. Martin continued to promote and sell globes for another twenty years or so, and three names – Senex, Ferguson and Martin (with a few rare exceptions) – appeared on the globes throughout this time. At some point (predating Captain Cook's voyages, as they are not shown) Martin issued a completely new terrestrial globe, illustrated here. It was engraved by Thomas Bowen (*c.* 1733–1790), who had collaborated with Martin on several projects. The revisions include more detail of recent discoveries around the North Pole and George Anson's track of 1744. There is also now a paper scale giving the distance of the sun from the equator (declination) throughout the year pasted on to the surface in the Pacific Ocean, an hour circle printed around the circle of 70°N latitude, and the prime meridian marked as passing specifically through Greenwich (rather than London), possibly for the first time on a globe.

*overleaf*
The terrestrial North Pole.

# 30

## CELESTIAL GLOBE, 1757/1770

John Senex and Benjamin Martin
Engraved, 30.5 cm (12 in.)
British Library, London

The cartouche on the celestial globe states 'A New Caelestial GLOBE wheron ye Stars are Carefully laid down from ye Correct Observations of Mr. Hevelius, Capt. Halley &c. By Ion. Senex F.R.S. Now made & sold (with very considerable improvements) By B. Martin in Fleet Street'. In fact the globe, which had first been issued shortly after 1710, hardly changed during the course of its existence (a grid was added to the zodiacal band). It was to become obviously outdated when George Adams, a rival to Martin, published his own new globes in 1766 (see Globes 33 and 34). Adams's celestial globes were very up to date and used the star catalogues of John Flamsteed published posthumously in 1725, and Nicolas de Lacaille (1713–1762), whose new observations in the southern hemisphere had been published in the 1750s. Martin considered Adams's globes a serious threat to his own, and in his *Appendix to the Description and Use of the Globes* of 1766, he published a scathing attack on them, ridiculing the details of their mounting, calling new geographical data 'confusing' and the new constellations 'unnecessary'. In spite of the threat posed by Adams, Martin's globes continued to sell for many more years, perhaps because they were significantly cheaper.

*overleaf*
The celestial polar area.

# 31

## TERRESTRIAL GLOBE, 1766

Anders Åkerman
Engraved, 59 cm (23 in.)
Royal Swedish Academy of Sciences, Stockholm

In 1758 the University of Uppsala in Sweden founded the Cosmographical Society in order to broaden horizons in geography and astronomy. Of course, maps and globes were essential to this aim. Up until then globes had to be imported from abroad, which was prohibitively expensive, and so very few globes had found their way to Sweden. Therefore the Society promoted the idea of making Swedish globes and maps. The engraver and mathematician Anders Åkerman (*c.* 1722–1778), a founding member of the Society, was chosen for the task of making globes. He completed his first pair with a diameter of 30 centimetres (12 inches) in 1759 and went on to produce globes 11 and 59 centimetres (4¼ and 23 inches) in diameter. The geography of the terrestrial globe was based on the latest maps acquired by Uppsala University and the Cosmographical Society. They included the *Atlas Russicus*, the first complete printed map of Russia, published in 1745, and the recent discoveries of Vitus Bering (1681–1741) along the north-east coast of Asia. The prime meridian passes through Ferro (El Hierro), the most southerly and westerly of the Canary Islands, perhaps because it did so in much of Åkerman's source material. The language of the globe is Latin.

A truly innovative feature of Åkerman's terrestrial globes is that they show lines of magnetic declination or variation of the compass. There had been increasing interest throughout the eighteenth century in this phenomenon, after Edmond Halley (1656–1742) published his magnetic map of the globe in 1701. For his lines, Åkerman used information on a chart compiled by Johan Gustav Zegollström in 1755, which was based on journals from Swedish East India ships.

*overleaf*
Lines of magnetic declination
in the Atlantic Ocean.

# 32

## CELESTIAL GLOBE, 1766

Anders Åkerman
Engraved, 59 cm (23 in.)
Royal Swedish Academy of Sciences, Stockholm

The celestial globe is based on the star catalogues of John Flamsteed (1648–1719) and the French astronomer Nicolas de Lacaille and is accurate for the epoch 1760. Fifteen new constellations appear on the globe. In the southern hemisphere, fourteen new constellations are depicted. On a four-year expedition to the Cape of Good Hope organised by the Académie Royale des Sciences, Lacaille had charted the stars in the southern hemisphere and had devised these constellations. Being a man of the Enlightenment, he had based them on the instruments and tools of the arts and sciences, rather than animals and mythical figures as in the past. Åkerman had included them on his celestial globe of 1759 and was one of the first globe-makers to do so. In this he beat George Adams, the renowned London instrument-maker who supplied many scientific instruments to Sweden, by seven years. In the northern hemisphere, one new constellation, Reno (the reindeer) appears on the globe. This constellation was codified by another French astronomer, Pierre Charles Lemonnier (1715–1799), in 1736, when on an expedition to Lapland to measure the arc of a meridian in order to test Isaac Newton's belief (which proved to be correct) that the earth is flattened at the poles. The expedition was organised by Professor Anders Celsius (1701–1744) of Uppsala University, who also developed the temperature scale that bears his name. The colouring is muted compared with other contemporary globes, the shading being restricted to pale greens and yellows. The gores meet at the equinoctial poles rather than the ecliptic poles. Reno did not become an established constellation.

*overleaf*
The constellation of Reno
(the reindeer).

# 33

## TERRESTRIAL GLOBE, c. 1766

George Adams
Engraved, 46 cm (18 in.)
British Library, London

In 1766 George Adams (1709–1772), mathematical instrument-maker to King George III, published a treatise on the use of globes to accompany two new pairs of globes he had made, with diameters of 30.5 and 46 centimetres (12 and 18 inches). Adams announced in his treatise that these globes were 'of a construction new and peculiar, being contrived to solve the various phenomena of the earth and heavens, in a more easy and natural manner than any hitherto published'. Though at first glance the terrestrial globe looks similar to any other, Adams had designed his globe to represent the 'real' motion of the earth so that, when solving problems with the globe, it should be rotated from west to east, reflecting the rotation of the earth (while the heavens appear to rotate from east to west). In order to signify the difference between this globe and others, the latitude scale on the brass meridian ring faces the west point on the horizon ring, whereas on conventional globes it faces east. In practice this was little more than a gimmick and made no difference to the way the globe could be used, as all the usual features of a globe were still in place.

As this terrestrial globe was published before Captain Cook's voyages to the southern seas, very little is shown of the coastline of New Holland (Australia). Van Dieman's Land (Tasmania) is still shown as being part of the larger mass of land, and New Zealand is depicted as nothing more than a few short indications of coastlines. Upon Cook's return to England in 1771, these shortcomings were quickly corrected, and the new discoveries were shown on updated globes by Adams and others.

*overleaf*
Hollandia Nova (Australia) and Nova
Zelandia (New Zealand) depicted on
Adams's globe of 1769.

*pages 148–9*
Hollandia Nova (Australia) and Nova
Zelandia (New Zealand) depicted on
Adams's updated globe of 1772.

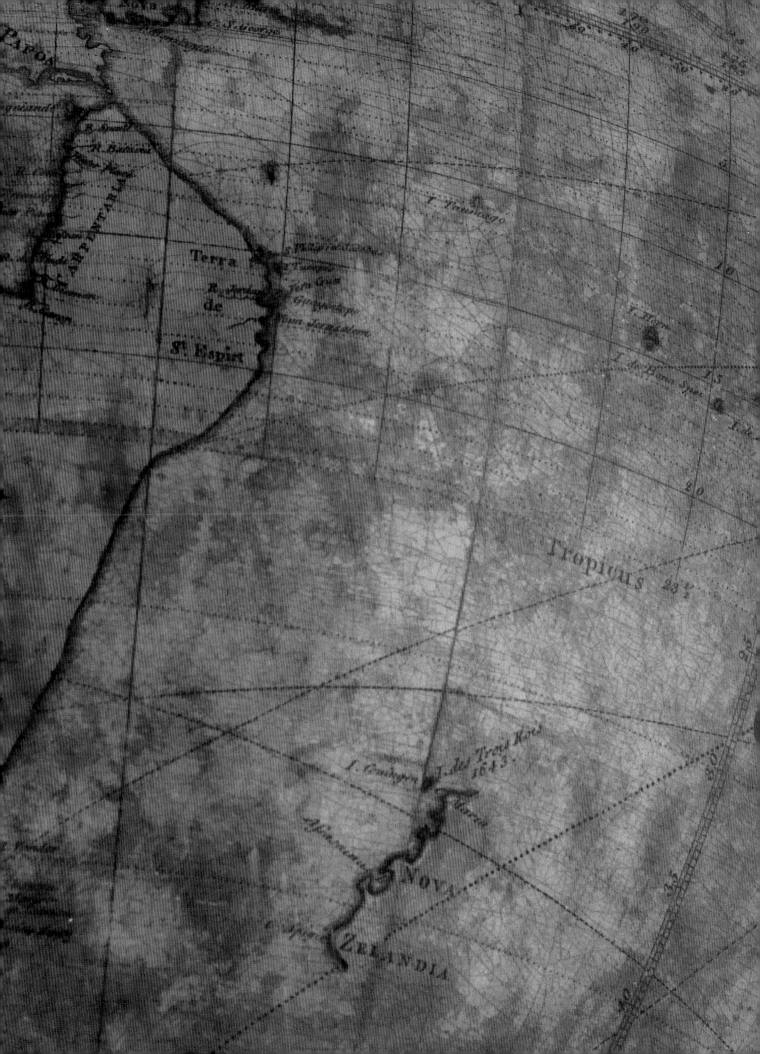

# 34

## CELESTIAL GLOBE, 1772

George Adams
Engraved, 46 cm (18 in.)
British Library, London

The first edition of Adams's celestial globe was right up to date, depicting the latest astronomical discoveries. In 1763, three years before the appearance of the globes, the final work of the French astronomer Nicolas de Lacaille was published posthumously. Lacaille had created fourteen new constellations (see Globe 32) and Adams depicts these on an English globe for the first time. Therefore, amidst the older constellations of animals and mythical figures, we can see the Sculptor's Workshop, the Painter's Easel, the Gravers, the Clock, the Chemical Furnace, the Air Pump, Hadley's Octant, the Sextant, the Telescope, the Microscope, the Mariner's Compass, Euclid's Quadrant and the Rhomboidal Net. The fourteenth constellation is named after the Table Mountain, from where Lacaille carried out his observations. The drawings of the Microscope and Air Pump seen on the globe are in fact based on the instruments that Adams himself made and sold in his shop. He also used the star catalogue by John Flamsteed published in 1725. In addition to these new constellations, and to give his globes yet another superior feature, Adams shows the 'mansions of the moon of the Arabian Astronomers', which as he states in his treatise 'will serve to show, how the moon passes from star to star, in the course of one or several nights, which is a very curious and useful amusement'.

Whereas the terrestrial globe was quickly updated to show new discoveries, this celestial globe of 1772 remained the same as earlier editions, by Adams, for there were no changes to the constellations.

*overleaf*
Two of the new constellations
formed by Nicolas de Lacaille:
Telescopium and Microscopium.

# 35

## POCKET GLOBE, 1779

Nicolas Lane
Engraved, 7 cm (2¾ in.)
British Library, London

In around 1779, Nicholas Lane (active 1775–83), about whom little is known, published a pocket globe, an example of which is shown here. He must have acquired the celestial gore plates made by Richard Cushee earlier on in the century, for they are exactly the same, with the heavens appearing in concave form, as we see them from the earth. The terrestrial globe, however, was completely new. On this particular globe, below the cartouche, the date 1776 and the name Prockter (the engraver of the globe) can be seen, but James Cook's final voyage of 1776–9 and the place of his death, 'Owhyhee', is also marked, so the globe must have been issued after this event. Lord George Anson (1697–1762) successfully circumnavigated the world and captured a Spanish treasure ship. He returned to England in 1744 and the track of his ship is shown. California is depicted as a peninsula rather than an island. The east coast of New Holland (Australia) is shown, but Dimens Land (Tasmania) is still connected to New Holland. The North and South Islands of New Zealand are also depicted.

In 1807, almost thirty years after this globe appeared, Lane's son Thomas (active 1801–29) reissued the globe with some revisions. Many more revisions were made in the following years and the globe sometimes appears with the labels of other sellers – for example, Schmalcalder and Bleuler – pasted over the name Lane. The globe was still being sold in the 1830s (see Globe 45, p. 182).

# 36

## TERRESTRIAL GLOBE, 1783

Gabriel Wright and William Bardin
Engraved, 23 cm (9 in.)
British Library, London

The globes made by Gabriel Wright (active 1782–1803) and his collaborator William Bardin (active 1775–98) marked a new phase in English globe-making that continued well into the nineteenth century. By the 1780s the globes of previous makers had been in existence for quite some time, and even with revisions they were looking outdated. It is not clear how Wright and Bardin came to work together, but Wright had been employed by Benjamin Martin (see Globes 29 and 30, pp. 129, 132) for many years, and in addition to making scientific instruments, he knew how globes were produced. It is likely that he felt the time was right for completely new globes to be published.

The first globes they made, with diameters of 23 and 30.5 centimetres (9 and 12 inches), were published in 1782. There are several versions of these globes, with different wording on the label in the North Pacific Ocean, though the cartography appears to be the same. The labels claim that the globes are improved, and in a book Wright published in 1783 he explains how. His innovation was to print hour circles onto the globes around the poles, to print the hours around the equator, and to place small brass pointers between the globe surface and the meridian ring. This allowed the globes, without the hindrance of a brass hour circle attached to the meridian ring, to be inverted within their stands so that 'the new discoveries, tracks, &c. may be clearly traced by the eye over all parts of the globe in a manner more conspicuous and easy than maps will admit of.'

*overleaf*
Label linking the globe with the
*Geographical Magazine* (see p. 161).

# 37

## CELESTIAL GLOBE, 1785

Gabriel Wright and William Bardin
Engraved, 23 cm (9 in.)
British Library, London

This globe, together with its terrestrial partner (Globe 36, p. 156), formed part of an interesting publicity strategy. A new monthly serial, *The Geographical Magazine; Or a New Copious, Compleat and Universal System of Geography*, was launched in January 1782. As a promotional inducement, subscribers were offered a pair of globes 'gratis'. The price for the magazine, published by Harrison & Co., was 2 shillings 6 pence per month. After buying the first twenty editions readers could acquire a terrestrial globe, and after the fortieth edition they could claim its celestial partner.

The collaboration of Wright and Bardin seems to have ended after these globes, but William Bardin and his successors went on to become one of the leading globe-making firms of the nineteenth century. In 1799 he published 'The New British Globes' with diameters of 30.5 and 46 centimetres (12 and 18 inches). The Wright/ Bardin globes, and later Bardin's globes, were also sold by the firm of instrument-makers and sellers, W. & S. Jones, whose labels can often be seen on the globes.

# 38

## TERRESTRIAL GLOBE, 1785

Miss Cowley
Engraved, 9 cm (3½ in.)
British Library, London

In the second half of the eighteenth century the teaching of geography to children became much more widespread. Self-assembly paper globes such as this example were made as inexpensive educational aids for children. In addition to teaching geography, the globes aimed to encourage the development of manual skills and to provide some fun. This globe has fourteen sections but larger ones were also made. A booklet accompanied the globe, with instructions for assembly and explanations of its various circles and features. Around the outside edge of the uppermost circle it is stated that this globe was invented by Miss Cowley and that it was published by John Marshall (1756–1823), a printer, bookseller and map-seller who specialised in producing material 'for the instruction and amusement of young minds'. Similar globes continued to be made in the nineteenth century.

# 39

## TERRESTRIAL GLOBE, 1790

Giovanni Maria Cassini
Engraved, 33 cm (13 in.)
National Maritime Museum, Greenwich

After Vincenzo Coronelli's outstanding achievements in the late seventeenth century, no new globes were made in Italy for almost a hundred years. Giovanni Maria Cassini (1745–*c.* 1824; no relation to Giovanni Domenico Cassini, founder of the Paris Observatory in 1671), was a highly skilled engraver who, when young, had worked for the artist Giambattista Piranesi (1720–1778), and he revived the art of globe-making in Rome. Cassini published this terrestrial globe in 1790, using the publications of the Académie Royale des Sciences in Paris and incorporating many of the most recent discoveries, relying particularly on the voyages of Captain Cook. All three of his voyages are shown, and much of the cartography is based on Cook's observations. Cassini was completely up to date with the politics of North America, naming the thirteen states along the eastern seaboard 'Provincie Unitie' (united provinces).

The language of the globe is Italian. The stand is simple, with the meridian ring made from wood, over which printed paper graduations are pasted. This reflects a general change in globe-making: new models were being produced at lower cost, and so were more affordable, satisfying a demand from a growing middle-class market. A celestial partner followed in 1792.

# 40

## GLOBE GORES, 1792

Giovanni Maria Cassini
Engraved, 33 cm (13 in.)
British Library, London

Cassini echoed Vincenzo Coronelli in several ways: like Coronelli, Cassini was a member of a religious order, the Ordine dei Clerici Regolari Somaschi, or CRS (which is stated in the cartouches on his globes), and he was extremely prolific in his output of maps, charts and views. He is probably best known for his large atlas, the *Nuovo Atlante Geografico Universale (New Universal Geographical Atlas)*, which was produced in three volumes between 1792 and 1801.

In the first volume, the twelve gores of each of his globes are printed together, with the horizon and meridian rings, which was an allusion to Coronelli's work, the *Libro dei Globi*, in which he published all of his globe gores. Cassini also included a chapter explaining how he constructed the gores for his globes, a rarity among globe-makers.

For the celestial globe, Cassini drew on the star catalogues of John Flamsteed and Nicolas de Lacaille. In addition, Cassini also included the two constellations formed by Pierre Charles Lemonnier, La Renna (the reindeer) and Solitario (a rare bird, called the Solitaire). These new constellations are depicted with broken lines to distinguish them from older ones.

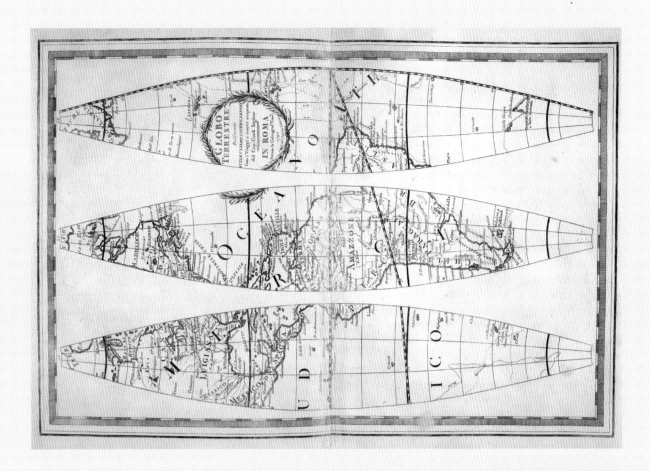

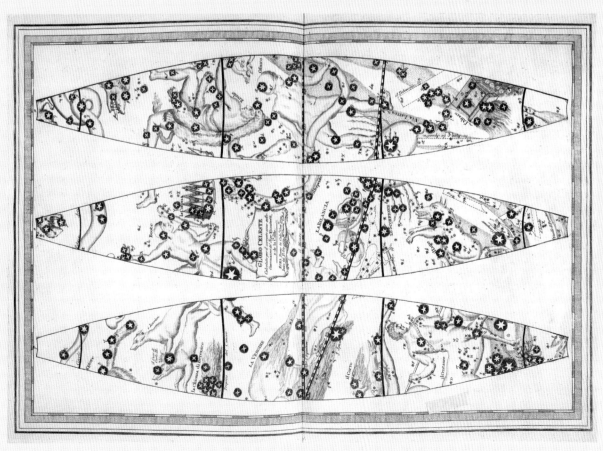

# 41

## TERRESTRIAL GLOBE, c. 1790

George Adams, Gabriel Wright and William Bardin
Engraved, 23 cm (9 in.)
British Library, London

In his lectures on astronomy and geography published in 1794, George Adams Junior (1750–1795) introduces the unusual mounting illustrated here as 'a new Apparatus adapted thereto, for solving, in an easy and natural manner, the several phenomena of the Sun, Moon, and Earth'. Made by Gabriel Wright and William Bardin (see Globes 36 and 37, pp. 156, 161), this globe is mounted at an angle of 23.5 degrees above a circular calendar plate which also represents the ecliptic circle. A movable arm below the calendar plate supports a vertical pillar, on top of which is a brass ball representing the sun. A spike extending from this denotes a ray of light. Two parallel vertical circles which pass around the globe represent the terminator (the limit of the sun's light) and the twilight zone.

This new mounting allowed various phenomena relating to the sun to be seen more clearly than with a standard globe, as the obstructions that the horizon ring and meridian inevitably create were removed. It was a forerunner of the modern globe, mounted at an angle and without a horizon ring. The only other examples of globes mounted in a similar manner are a terrestrial and celestial pair 46 centimetres (18 inches) in diameter ordered in 1790 by Martinus van Marum specifically for the Teyler Foundation (now Teyler's Museum) in Haarlem, which was building up a collection of scientific instruments to form a Physics Cabinet, to be used for demonstration and teaching.

# 42

## POCKET GLOBE, 1793

John Miller
Engraved, 8 cm (3 in.)
British Library, London

Although there are records of globe-making attempts in Scotland earlier than 1793, the date of this example by John Miller (1746–1815), no earlier globe survives. Miller, who was born in Edinburgh, had worked for George Adams Senior, a prominent instrument-maker in London, in the 1760s. He returned to his native town in 1769, and his main employment was supplying instruments to land surveyors. In 1776, he and a partner, John Ainslie (1745–1828), a surveyor and engraver, embarked on making a 30.5-centimetre (12-inch) globe pair, to be financed by subscription. Unfortunately not enough subscribers came forward and the project did not come to fruition.

Seventeen years later, in December 1793, Miller advertised a pair of 9-inch (23-centimetre) globes, no examples of which are known to survive, and a 7.5-centimetre (3-inch) pocket globe that was advertised as a 'New Year's Gift for the instruction and amusement of Young Ladies and Gentlemen' at half a guinea. Several examples of the pocket globe are extant. The terrestrial globe is up to date, with Cook's third voyage and death noted, though in comparison with other contemporary pocket globes, the amount of geographical information is sparse. The celestial gores lining the case are concave, showing the heavens as they would be seen from earth. The language of the terrestrial globe is English; the celestial globe is a mixture of Latin and English.

# 43

## SELENOGRAPHIA, A LUNAR GLOBE, 1797

John Russell
Stipple-engraved, 30.5 cm (12 in.)
British Library, London

John Russell RA (1745–1806), painter to George III, had a passion for the moon, which he observed, through a telescope, in minute detail. A fashionable artist well known for his pastel portraits, he used his skills to describe the moon's visible surface in many pastel drawings and pencil sketches. The culmination of his study was his selenographia, a stipple-engraved lunar globe attached to an instrument, designed by himself, 'contrived to give it such motions as will exhibit all the appearances which the face of the moon puts on to the inhabitants of the earth'. The ensemble can give a demonstration of the moon's libration, the apparent oscillation by which different areas at the edge of the visible moon sometimes come into view. The small terrestrial globe enables us to see the effects of the moon's parallax.

Russell's understanding of the visible surface of the moon was undoubtedly among the most advanced at the time. Among his many portrait subjects were the astronomer Sir William Herschel (1738–1822), the discoverer of the planet Uranus, and the Astronomer Royal, Nevil Maskelyne (1732–1811). After the invention of the telescope in the early seventeenth century, the moon could be studied in much greater detail, and it fascinated many people, including Christopher Wren (1632–1723), who made a relief moon globe in 1661 which has not survived, and Tobias Mayer (1723–1762), an astronomer in Göttingen, who had plans to make a printed lunar globe in the 1750s. Mayer got as far as having copper plates engraved, but financial problems prevented the manufacturing of any globes. John Russell's globe is the earliest surviving printed lunar globe.

*above*
The small ball at the front is a terrestrial globe.
The selenographia gives an expanded view
of the moon from the earth.

*opposite*
The mechanism of the selenographia.

# 44

## TERRESTRIAL GLOBE, 1800

John Cary
Engraved, 53.5 cm (21 in.)
British Library, London

John Cary (1755–1835), engraver and publisher, established a family firm of map-makers in the late eighteenth century that went on to be one of the most successful and prolific of the nineteenth century. Cary globes were first advertised in the *Traveller's Companion* of 1 January 1791 and were offered for sale in sizes of 3½, 9, 12 and 21 inches (9, 23, 30.5 and 53.5 centimetres) in diameter. Over the following years the firm made other sizes of 6, 15 and 18 inches (15, 38 and 46 centimetres). This is a much greater range than had been offered before, suggesting that the globe-buying public was growing larger. As with earlier globes, there was a choice of stands to suit different tastes and purses. The 9-centimetre (3½-inch) globe could be bought as a pocket globe or, at slightly greater expense, mounted in a stand.

The globe illustrated here, published in 1800, was current for its time, showing the tracks and discoveries along the north-west coast of America by George Vancouver (1758–1798), who had returned to England in 1795 after a four-and-a-half-year voyage. We also see the track of Jean François de la Perouse (1741–*c.* 1788) off the north-east Russian coast. Cary was keen to update his terrestrial globes as new information became available, and the date of the latest additions can often be seen engraved below the cartouche. At first Cary made his globes in collaboration with his brother William (1759–1825), a scientific instrument-maker, and his sons John and George joined the business in the 1820s. John Cary Junior died in 1852 but the firm continued under different management, using the name 'William Cary', until the 1890s.

# 45

## POCKET GLOBE, 1819

Thomas Lane
Engraved, 7 cm (2¾ in.)
British Library, London

The terrestrial pocket globe first issued around 1779 by Nicholas Lane (see Globe 35, p. 155) underwent many revisions. In the 1807 update, issued by his son Thomas (active 1801–29), New South Wales, Botany Bay and Cape Byron are depicted in New Holland (Australia) and 'Buenos Ayres' (Buenos Aires) appears in South America. Two years later there are more changes: Dimens Land (Tasmania) is separated from New Holland by the Bass Strait; Port Jackson (Sydney) is added to the eastern coast of the mainland; and Sharks' Bay and 'South C.' are newly marked on the western side. The Antipodes of London are also shown. In north-west America, New Albion and the Stony Mountains (the Rockies) have been added. Curiously, the date of Captain Cook's death, 14 February 1779, is another late addition squeezed in below the Sandwich Islands.

By 1819, the date of the globe shown here, there are more additions. At the southern tip of the Californian peninsula, C. S. Lucas (Cape San Lucas) is now shown, and an area of 'Gold Mines' is located in South America. Dampier's Anchor, where William Dampier first reached Australia, is marked off the north-west coast of New Holland, and we can see a mysterious 'Labyrinth' off the north-east coast. The celestial gores of the Lane pocket globes issued in the nineteenth century are usually coloured a characteristic green, sometimes with a paler band to show the Milky Way, rather than having individually coloured constellations. Lane also acquired the plates for a 7.5-centimetre (3-inch) pocket globe, possibly from the instrument-maker Dudley Adams, and so the early nineteenth century saw both 7- and 7.5-centimetre (2¾- and 3-inch) globes bearing Lane's name.

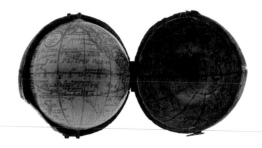

# 46

## TERRESTRIAL GLOBE, 1827

Didier Robert de Vaugondy/Charles-François and Félix Delamarche
Engraved, 22 cm (8½ in.)
British Library, London

Didier Robert de Vaugondy (1723–1786) came from a highly respected family of map-makers based in Paris. His first globe, made in 1745 with a diameter of 15 centimetres (6 inches), was dedicated to King Louis XV. This led to his prestigious appointment as Geographer to the King. He was then commissioned to make a pair of larger globes of 46 centimetres (18 inches) diameter for the French navy, though it is doubtful whether they were ever used as practical aids. Vaugondy made four sizes of globe ranging from 7.5 to 46 centimetres (3 to 18 inches) and the cost varied depending on how they were mounted. At the luxury end of his output, some of the stands were very ornate with a great deal of gold decoration.

Vaugondy wrote the entries for 'geography' and 'globes' in Denis Diderot's great *Encyclopédie* published in 1757. His firm was taken over by the publisher Jean Fortin (1750–1831) in the 1770s and was in turn acquired by Charles-François Delamarche (1740–1817), who made a great commercial success of the business. Delamarche aimed much of his production at the general public, publishing smaller globes in inexpensive mountings often made from pasteboard. These brightly coloured globes with their distinctive red edges were very decorative and much cheaper to buy than their predecessors. The firm continued to produce globes well in to the nineteenth century. This example was made by Félix Delamarche, the son of Charles-François, though the cartouche clearly states 'Successr. de Robert de Vaugondy'.

# 47

## TERRESTRIAL GLOBE 'THE EARTH & ITZ INHABITANTS', c. 1830

Lithographed, 5 cm (2 in.)
British Library, London

This small cardboard box contains a tiny globe and a strip of concertinaed paper that unfolds to reveal twenty-eight hand-coloured figures of the inhabitants of the earth dressed in their national costume. Though we find no maker's name on the box or globe, this charming novelty item was probably made for the foreign market by the Nuremberg-based family firm Bauer, which specialised in small educational globes. The identity of each figure, all men, is given in English, French and German. Similar globes can be found which are signed 'C.B.', that is, Carl Bauer (1780–1857).

The globe contains minimal information: only the names of the continents and a few countries are given. The language of the globe is English. Tierra del Fuego is translated literally to become Fireland. The label on the box lid, with the misspelling of the word 'its', adds to its charm.

# 48

## TERRESTRIAL GLOBE, 1831

Newton & Berry
Engraved, 4 cm (1½ in.)
British Library, London

John Newton (1759–1844), the founding member of this family firm, started making globes in the 1780s. He was joined at different times by various members of his family and, in the 1830s, by Miles Berry (*c.* 1803–1843). After Newton's death the firm carried on through successive generations. The firm advertised globes in a remarkable range of ten sizes from 1 to 25 inches (2.5 to 63.5 centimetres) in diameter, and buyers could choose between stands. The firm also made planetaria, mechanical models of the planetary system, in various sizes.

This terrestrial globe is an example of the smallest made by the firm. Given its tiny size, the amount of information shown is astonishing. The prime meridian passes through London. We see the track of Captain Cook's final voyage of 1776 and also that of Captain Charles Clarke and Lieutenant John Gore. Cook was killed by Hawaiian natives in the Sandwich Isles (Hawaii) on 14 February 1779, and Clarke took over the ill-fated expedition, appointing Gore second-in-command. As detailed on the globe, they sailed north to the Bering Sea and then started the long return journey back to England. Captain Clarke died of consumption shortly after the return voyage began. Under Gore's command, the expedition finally reached England again in October 1780.

# 49

## TERRESTRIAL GLOBE c. 1834

Pendleton's Lithography
Lithographed, 14 cm (5½ in.)
British Library, London

This rather strange-looking globe is more significant than its appearance might indicate. A small cartouche in the south Atlantic Ocean states 'Pendleton's Lithy. BOSTON'. The lithographic printing process was introduced into America around 1819, almost two decades after it took off in Europe. In the 1820s several lithographic printing establishments were set up in Boston, New York, Philadelphia and Washington. William Pendleton (1795–1879) originally trained to be an engraver. He and his younger brother John, while travelling in Europe, had learned about the new printing technique of lithography. The firm founded by the brothers in Boston in 1825 was more successful than many others and produced a wide range of lithographically printed material, which included portraits, reproductions of paintings, maps and town plans, though William continued to use his engraving skills.

The Pendletons had connections with other printers such as William Annin (active 1813–39), who engraved globes by Josiah Loring (1775–1840), and other contemporary printed globes might have inspired their effort. The globe illustrated here appears to be one of the first lithographically printed globes in America. It has twelve gores and two polar caps pasted onto a solid wooden ball which has shrunk and split, demonstrating the impracticality of using solid wood. The paper is now discoloured because of its direct contact with the wood, though five colours can still be made out. The intended purpose of this globe is unclear. Very little information is shown, so it has no great value as a geographical tool. Few examples of this globe have survived, and it may have been an experiment.

# 50

## TERRESTRIAL GLOBE, 1837

James Wilson/Cyrus Lancaster
Engraved, 33 cm (13 in.)
Sterling Memorial Library Map Department, Yale University

Globes were expensive to buy and import, so it is not surprising that entrepreneurs attempted to produce them on American soil. There are records of globes being made in late eighteenth-century America, but the first successful commercial globe-maker there was James Wilson (1763–1855) of Vermont. Originally a farmer and a man of many talents, Wilson taught himself to make globes from scratch after becoming fascinated with a pair he had seen at Dartmouth College in Hanover, New Hampshire. The college, perhaps surprisingly, owned three pairs of globes, one of which was known to be by John Senex. For his own globes Wilson did everything himself. He made the spheres and turned the wooden stands, and used his skill as a blacksmith to make the brass meridian rings and quadrants. Wilson consulted Amos Doolittle (1754–1832), a notable engraver, to teach him how to engrave on copper. He also sought advice from Jedediah Morse (1761–1826), author of *Geography Made Easy* (first published in 1784) on the matter of plotting maps and gores, which he printed himself. He also made his own inks, glues and varnishes.

Commercial production started around 1810, and the price of a pair of globes was $50. Wilson's globes were cheaper than imported ones, as well as being more accurate in their depiction of American lands and territories.

# 51

## CELESTIAL GLOBE, 1837

James Wilson/Cyrus Lancaster
Engraved, 33 cm (13 in.)
Sterling Memorial Library Map Department, Yale University

Wilson ensured that his celestial globes were up to date, and he employed the latest star catalogues. His globes were made in three sizes – 7.5, 23 and 33 centimetres (3, 9 and 13 inches).

The firm became successful and a factory was established in Albany, New York, in about 1815. It continued, with the help of Wilson's sons John and Samuel, and later another son, David, until 1833, when Wilson's employee and son-in-law Cyrus Lancaster took over sole management of the company.

As with many other globes, an instruction manual on their use was published. This was based on the manual by the Englishman Thomas Keith (*c.* 1759–1824), whose very popular treatise on globe use ran to at least twenty-five editions after its publication in 1805.

While it is easy to take the pairing of globes for granted, it is worth noting that Wilson felt it appropriate to match his terrestrial globe with a celestial example. By the nineteenth century it is far from obvious that this arrangement has any practical or educational value beyond the information presented by each globe individually. The need to provide a *pair* of globes arose within the sixteenth-century discipline of cosmography, where the earth and the heavens were studied as an integrated whole. The resultant pairing of terrestrial and celestial globes had become such an established convention that it survived translation in time and place to nineteenth-century America.

# 52

## CELESTIAL GLOBE, 1845

Bale
Engraved, 13 cm (5 in.)
British Library, London

In the mid-nineteenth century this type of mounting – where the globe is permanently suspended at an angle of 23.5 degrees in a semi-meridian ring on a single pillar – started to be more common, gradually taking over from what had been the standard mount since the time of Mercator. Gone are the horizon ring and the adjustable meridian ring, which had allowed complicated problems to be solved. An advantage of this simpler type of mounting is that more of the globe can be easily seen, especially in the southern hemisphere.

We know very little about Bale (active 1845–6), other than that he made a celestial globe in 1845 and joined forces with George Woodward (active 1820–55) to produce pairs of globes. They worked in London and together they published globes of 8, 13 and 18 centimetres (3, 5 and 7 inches).

# 53

## TERRESTRIAL GLOBE, 1846

George Woodward
Engraved, 13 cm (5 in.)
British Library, London

Although this terrestrial globe is small, it holds a considerable amount of information. There are no tracks on this globe, but it does show the recent discoveries of John Biscoe (1794–1843), who circumnavigated Antarctica in 1831 and discovered Enderby Land, Adelaide Island, Graham Land and the Biscoe Islands. The most recent discovery depicted is that of the whaling captain John Balleny, who in 1839 discovered a small group of islands in the region that were named after him. The islands in the South Pacific are shown in great detail and are described as 'Dangeros Arch' (Dangerous Archipelago).

An analemma is present in the Pacific Ocean, but this globe and its celestial partner are more cartographical tools than scientific instruments, and reflect a wider interest in geography and a growing demand for cheaper globes.

*overleaf*
Islands in the South Pacific.

C E A N

Woahoo
Mowee
Cowhyee

Palmyras
Washingtone       Low Ts
Fannings
Christmas        150
160                    140
XIV

Browns
Malden   Vohotir                    Marques
Starbede                                        Declination
A                                              for every

Penhryns        Caroline        Madalena
Caroline                          Low Ts
Society Islands    Keoll          Heden

Savaroffs                                      Minerva
20   Uliea      Dean                          Whitsunday
Palmerston   Harveys   S.Pablo  Arch.
Watechoo   Onaheite   Dangroo   Margarete
Roxburgh            Resolution  Toobonai  Vandriver
Opero   Crowns of Thorn

S

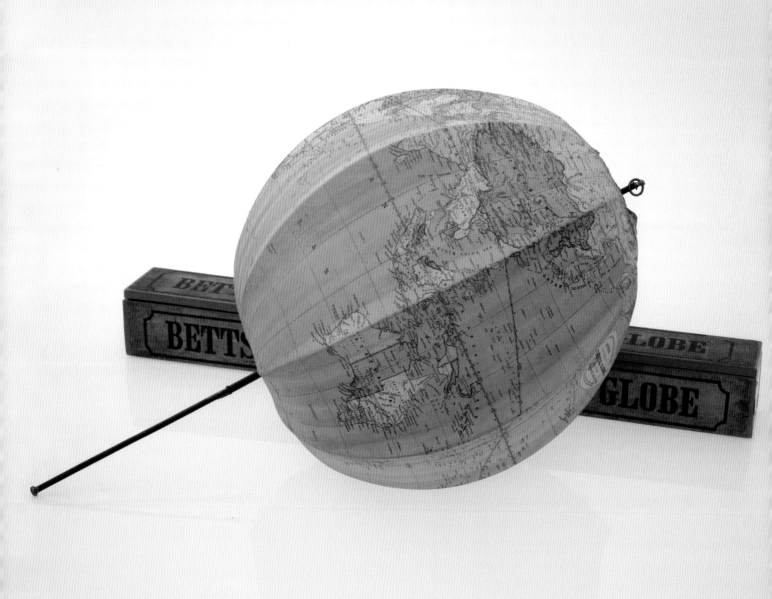

# 54

## TERRESTRIAL UMBRELLA GLOBE, c. 1860

John Betts
Colour lithographed, 40.5 cm (16 in.)
British Library, London

A globe, even one of modest diameter, can be cumbersome and difficult to move about when placed in a stand. It is no surprise, then, that some thought was given to imaginative and ingenious designs for making globes more portable. John Betts (active 1844–75), who made a variety of educational aids for the teaching of geography, developed two solutions. His first portable globe, published about 1850, with a diameter of 13 centimetres (5 inches), comprised eight cardboard gores, linked together by string drawn through an ivory toggle. When the toggles were pulled together, the gores expanded to form an approximation of a sphere. Betts's 'umbrella' globe, for which he is best known and which is illustrated here, comprises eight gores of linen, sewn together and mounted onto a metal frame with struts, with an umbrella-like mechanism. Though these globes have a novelty value, their geographical content is accurate and they show some of the most recent discoveries of the time.

# 55

## TERRESTRIAL GLOBE, 1861

Dietrich Reimer
Colour lithographed, 78 cm (31 in.)
National Maritime Museum, Greenwich

Dietrich Reimer's (1818–1899) publishing house, founded in 1845 in Berlin, concentrated on maps, geography, archaeology and art. In 1852 he took over another company, Adami & Co, which had been making globes for some time, and added its globes to his own output. Reimer's company was very successful in all areas of activity and continued to make globes until the second half of the twentieth century. The range produced was large: from small, simple, single-pedestal globes for the schoolroom to more elaborate examples such as the one pictured here, on which a great amount of information is visible. Rather than tracks, we see major shipping routes. The recent discoveries in the area of the Antarctic are marked, including the magnetic South Pole, located by James Clark Ross in 1841.

Unusually, the globe features two prime meridians, one passing through Greenwich and the other through the Canary Islands. Corresponding scales of longitude are given above and below the equator. At this time there was still no international prime meridian used throughout the world and it was not until the International Meridian Conference of 1884 in Washington, D.C., that it was agreed that Greenwich would be the prime meridian of the world. This decision had far-reaching significance for it introduced a single time system throughout the world based on Greenwich Mean Time.

# 56 AND 57

## TERRESTRIAL GLOBES

Abraham Nathan Myers
Dissected globe, *c.* 1866
Colour lithographed, 17 cm (6½ in.)
Cardboard cut-out globe, *c.* 1875
Colour lithographed, 22 cm (8½ in.)
British Library, London

Abraham Nathan Myers (1804–1882) sold a range of fancy goods and toys, including geographical items such as these unusual globes. The dissected globe is cut into eight cross-sections which are also cut into segments. Maps are printed on the upper layers of the cross-sections, and information about the continents and their inhabitants is printed on the other side. The whole ensemble is a three-dimensional jigsaw puzzle and was designed as an educational game and aid in the teaching of geography. By assembling the sections correctly, which encouraged the reading of the information they contained, the outer globe map was automatically assembled. Dissected map puzzles like these were popular, and they were made by several other artisans. Few complete examples survive as the pieces were easily lost.

The cardboard cut-out globe (overleaf) includes an elaborate paper stand. It was an inexpensive way for children to acquire their own globe.

# Cardboard Designs

### for constructing a

### Terrestrial     Globe.

LONDON: A.N. MYERS & Cº

*Basis.*

*Foot for the Globe.*

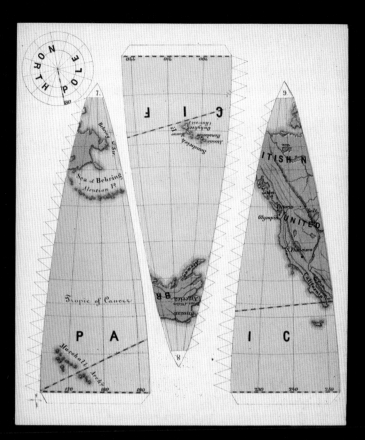

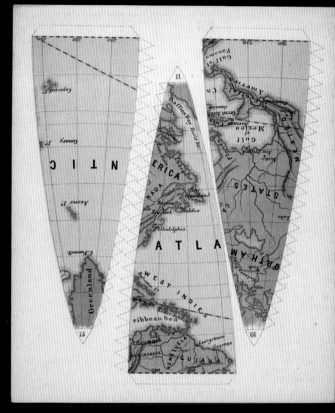

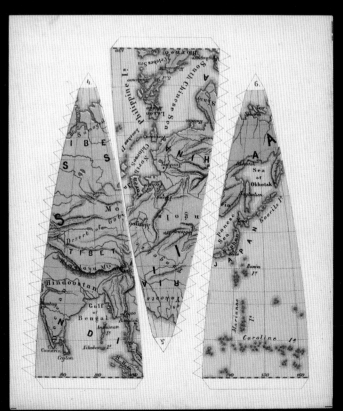

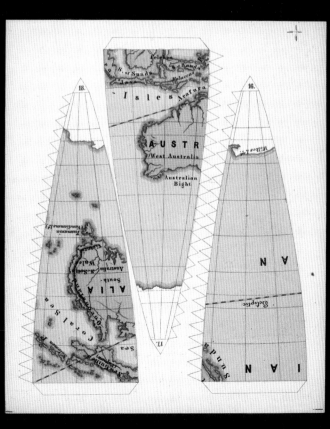

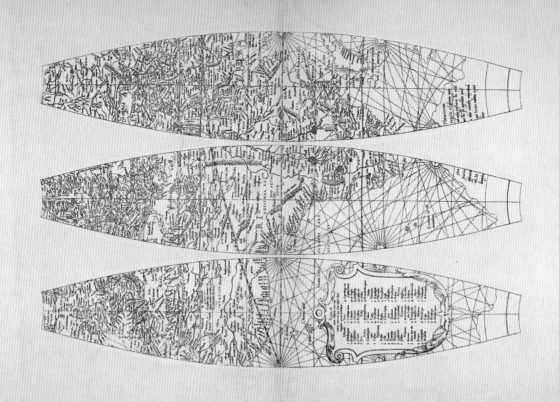

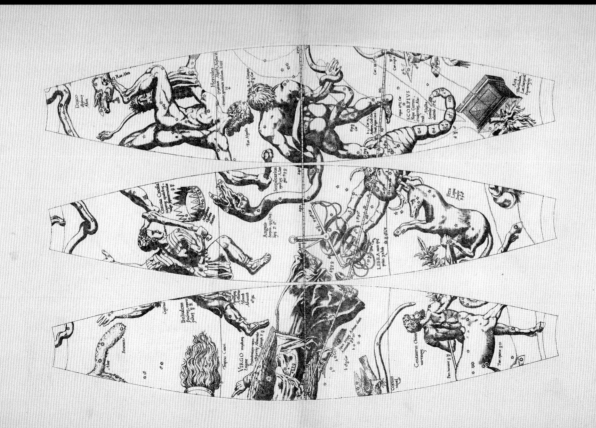

# 58

## FACSIMILE TERRESTRIAL AND CELESTIAL GORES, 1875

Gerard Mercator
Photolithographed
42 cm (16½ in.)
British Library, London

The discovery in 1868 of a set of original terrestrial and celestial gores by Gerard Mercator, during the dispersal of a private library after the owner's death, was greeted with great excitement by contemporary scholars, renewing interest in the work of early globe-makers. The gores were acquired for the Belgian nation by the Royal Library in Brussels (Mercator was of Flemish origin). It seems that at the time the work of the early globe-makers had been largely forgotten. Those involved with the find thought that these were the only gores of Mercator's globes in existence, which fostered a realisation of the value and importance of these fragile objects. Through the gores, Mercator's globes received a fresh look, and thus made the vital transition from being artefacts of little use, since they were out of date, to being historic objects of huge interest. A short biography of Mercator written in the following year stimulated further interest in his life and work.

In order to ensure the preservation of the precious gores, it was decided, with the approval of the Belgian government, to make a facsimile reproduction using a photographic technique. Reproductions of the globes themselves were made using the facsimiles and exhibited at the International Congress of Geographical Science in Paris in 1875. Shortly after this, original examples of Mercator's globes started to emerge from obscurity. Today, antiquarians have identified about forty Mercator globes, many forming pairs. Five hundred years after his birth, Mercator is no longer forgotten and is instead highly revered as one of the most important cartographers of all time.

# 59

## TERRESTRIAL GLOBE, c. 1875

W. & A. K. Johnston
Colour lithographed, 8 cm (3 in.)
British Library, London

Globes in traditional mountings were still in production at the end of the nineteenth century, but the practice was waning, and the majority of globes were now presented in much simpler stands. The globe shown here, and the following example, were made by the same company within about ten years, and they graphically demonstrate the old and new styles of globe-making.

Both globes were printed using colour lithography, which had superseded engraving as the printing method most commonly used. They were made by the Edinburgh-based firm W. & A. K. Johnston, run by the brothers William (1802–1888) and Alexander Keith (1804–1871), who were prolific map-makers and publishers. They had been apprenticed to James Kirkwood (*c.* 1746–1827), another Scottish map- and globe-maker, before setting up their own operation in 1826. Their early work was produced by engraving and letterpress, but by the mid-1860s they had converted to lithography, which enabled a much larger output. At the Great Exhibition in London in 1851, the Johnstons exhibited a 76-centimetre (30-inch) globe, showing geological, meteorological and physical features of the earth, the first of its kind, for which they were awarded a Prize Medal. From then on they published a wide range of globes, ranging from 8 centimetres (3 inches) to the large 76-centimetre (30-inch) model.

# 60

## TERRESTRIAL GLOBE, AFTER 1884

W. & A. K. Johnston
Colour lithographed, 15 cm (6 in.)
National Maritime Museum, Greenwich

This globe on its simple pillar stand makes a virtue of its low price: 'FIVE SHILLINGS' is printed on the label. New geographical features include Greely Fiord, named after the American polar explorer Adolphus Greely (1844–1935) who led an expedition to North Greenland from 1881 to 1884. In addition, rather than explorers' tracks, shipping routes – 'Steam Packet Route from Suez to Bombay' – and telegraph lines are shown.

After the deaths of the founders, their younger brother, Thomas Brumby Johnston (1814–1897), took over the firm, and his sons carried the business into the twentieth century. In addition to making globes for the home market, the firm exported school globes to America. The Chicago-based publisher of maps and globes, A. J. Nystrom and Co., which was founded in 1903, became the sole American agents for W. & A. K. Johnston. Later the globes were sold by other companies, but the copyright remained with the Johnston firm.

Compared with Martin Behaim's globe (p. 42), this humble example succinctly shows how globes had changed over the course of 400 years: having been precious, rare and novel objects with a limited circulation, they were now items of mass production available to everyone. The common thread is the fascination we all share in looking at a globe.

# GLOSSARY

**ANALEMMA**
In the context of a globe, this usually refers to a diagram showing the position of the sun at the mean solar noon over the course of a year, in the form of a figure-of-eight on a terrestrial globe.

**ANTIPODES**
Places on the surface of the earth directly opposite to each other.

**ASTERISM**
A group of stars smaller than a constellation.

**CALOTTE**
A circular paper cap pasted at the poles of a globe.

**CARTOUCHE**
A label on a globe, usually displaying the name of the maker, the date of manufacture and other descriptive text.

**CELESTIAL CO-ORDINATES**
There are two co-ordinate systems that can be used on a celestial globe. The first, following Ptolemy, uses the ecliptic as a baseline from which the position of the stars is measured using celestial longitude and celestial latitude. The second system is based on the celestial equator – a projection of the terrestrial equator. In this system star positions are measured by right ascension along the equator and declination above and below the equator. The vernal equinox or the First Point of Aries is the starting point for the measurement of celestial longitude and right ascension.

**CONSTELLATION**
A number of stars grouped together to form a recognisable imaginary figure.

**DECLINATION**
*See* celestial co-ordinates.

**COLURE**
One of four meridians which pass through the equinoctial points and the solstices.

**ECLIPTIC**
The great circle on the celestial sphere which is the apparent path of the sun through the stars. It cuts the equator at an angle of 23.5°. The ecliptic is divided into the twelve signs of the zodiac.

**ECLIPTIC POLE**
The north or south point at an angle of 90° to the ecliptic.

**ENGRAVING**
A method of printing where the lines that hold the ink are cut into a metal plate.

**EPOCH**
The year for which star positions on a globe have been calculated, taking precession into account.

**EQUATOR (TERRESTRIAL)**
The great circle on the earth which is equidistant from the North and South Pole and which divides the earth into the northern and southern hemispheres.

**EQUATOR (CELESTIAL)**
The great circle in the heavens which is a projection of the terrestrial equator.

**EQUATORIAL POLE**
The north or south pole at an angle of 90° to the equator.

**EQUINOX, EQUINOCTIAL POINTS**
The equinoxes are the two points where the ecliptic intersects the equator. At the vernal equinox the sun passes from south to north; at the autumnal equinox the sun passes from north to south.

**FUSEE**
A conical pulley with a helical groove around its surface, which receives a cord or chain as it winds on and off the mainspring barrel of a clock or watch, having the effect of equalising the force transmitted from the spring as it runs down.

**GORE**
One of the segments of paper that cover a globe. The use of twelve gores came to be standard practice, twelve being a very convenient factor of 360°.

**GREAT CIRCLE**
A circle on a globe, the plane of which passes through the centre of the globe.

**HORIZON RING**
The horizontal ring encircling the globe that represents the plane of the horizon on the earth. The horizon is a great circle that separates the visible half of the heavens from the invisible. On a globe, the printed horizon ring displays several concentric circles that typically show degrees of the compass, the points of the compass, the degrees and signs of the zodiac, and the calendar months and occasionally winds.

**HOUR CIRCLE**
A ring, placed around the pole, usually made of brass, and divided into twenty-four hours.

**LATITUDE**
The angular distance of a place on the earth's surface, north or south of the equator.

**LITHOGRAPHY**
A printing process which relies on the principle that oil and water repel each other and which requires no mechanical treatment of the printing plate. Originally the design was drawn onto a lithographic stone, but the process was adapted so that metal plates could be used.

**LONGITUDE**
The angular distance of a place on the earth's surface east or west of the prime meridian.

**LOXODROME**
*See* rhumb line.

**MAGNITUDE**
The apparent brightness of a star expressed as a number where one is the brightest. In pre-telescopic astronomy, six divisions were recognised.

**MAINSPRING**
A steel band wound up within a cylindrical barrel, which it turns as the band unwinds, to provide the driving force for a clock or watch.

**MANUSCRIPT GLOBE**
A globe where the cartographic information has been applied directly to the surface by hand, whether by drawing, paint, or, in the case of metal globes, by engraving.

**MERIDIAN**
A great circle which passes through the poles.

**MERIDIAN RING**
The brass ring in which a globe is suspended. It is usually engraved with degrees of latitude.

**NOVA**
A star showing a rapid increase in brightness, sometimes thought to be a new star as it becomes visible.

**ORRERY**
A mechanical model that shows the motions of the earth and moon (and sometimes also the planets) around the sun.

**PLANETARIUM**
A mechanical model that shows the movements of the planets.

**PRECESSION**
The slow westward motion of the equinoctial points along the ecliptic, resulting in the earlier occurrence of the equinoxes in each successive sidereal year, caused by the slow change of direction of the earth's axis of rotation.

**PRIME MERIDIAN**
The meridian where longitude is chosen to be 0° and from which longitude is measured.

**PRINTED GLOBE**
A globe where the cartographical information is printed onto paper gores which are pasted onto a sphere.

**QUADRANT OF ALTITUDE**
A thin strip of brass, carrying a scale of degrees, which is adjustably attached to the meridian ring and which enables distances to be measured.

**RIGHT ASCENSION**
See celestial co-ordinates.

**RHUMB LINE**
The course followed by a ship or other vessel sailing in a fixed direction.

**SIDEREAL**
Relating to the stars.

**SOLSTICE, SOLSTITIAL POINTS**
The points on the ecliptic at the maximum distance north or south of the equator.

**TRACK**
The route of a ship taken by seafarers.

**WOODCUT**
A method of printing which transfers ink from the relief areas of a carved block of wood.

**ZODIAC**
A band of the heavens 8° to either side of the ecliptic which is divided into twelve equal parts, each containing a sign or constellation of the zodiac.

# BIBLIOGRAPHY

Germaine Aujac, 'The Foundations of Theoretical Cartography in Archaic and Classical Greece' and 'Greek Cartography in the Early Roman World', in J. B. Harley and David Woodward (eds), *The History of Cartography, Volume One: Cartography in Prehistoric, Ancient, and Medieval Europe and the Mediterranean* (Chicago: University of Chicago Press, 1987), pp. 130–47, 161–76

Robert W. Baldwin, 'P. Giovanni Maria Cassini, C.R.S. (1745–ca.1824) and His Globes', *Der Globusfreund*, 43/44 (1995), 201–18

Peter Barber, 'Beyond Geography: Globes on Medals 1440–1998', *Der Globusfreund*, 47/48 (1999/2000), 53–80

Peter Barber and Tom Harper, *Magnificent Maps: Power, Propaganda and Art* (London: British Library, 2010)

A. D. Baynes-Cope, 'The Investigation of a Group of Globes', *Imago Mundi*, 33 (1981), 9–20

A. D. Baynes-Cope, *The Study and Conservation of Globes* (Vienna: Internationale Coronelli-Gesellschaft, 1985)

Silvio A. Bedini, *Thinkers and Tinkers: Early American Men of Science* (New York: Scribner, 1975)

H. von Bertele, *Globes and Spheres* (Lausanne: Scriptar, 1961)

Einar Bratt, *En Krönika om Svenska Glober* (Uppsala: Almqvist & Wiksell, 1968)

D. J. Bryden, 'Capital in the London Publishing Trade: James Moxon's Stock Disposal of 1698, a "Mathematical Lottery"', *The Library*, sixth series, 19 (4), 1997, 293–350

April Carlucci and Peter Barber (eds), *Lie of the Land: The Secret Life of Maps* (London: British Library, 2001)

Gloria Clifton, *Directory of British Scientific Instrument Makers 1550–1851* (London: Zwemmer/ National Maritime Museum, 1995)

Nicholas Crane, *Mercator: The Man who Mapped the Planet* (London: Weidenfeld & Nicolson, 2002)

Edward H. Dahl and Jean-François Gauvin, *Sphaerae Mundi: Early Globes at the Stewart Museum* (Montreal: Septentrion/McGill-Queen's University Press, 2000)

Elly Dekker, *Catalogue of Orbs, Spheres and Globes* (Florence: Istituto e Museo di Storia della Scienza, Giunti, 2004)

Elly Dekker, 'The Copernican Globe: A Delayed Conception', *Annals of Science* 53 (1996), 541–66

Elly Dekker, *Globes at Greenwich* (Oxford: Oxford University Press/National Maritime Museum, 1999)

Elly Dekker, 'Globes in Renaissance Europe', in David Woodward (ed.), *The History of Cartography, Volume Three: Cartography in the European Renaissance Part 1* (Chicago: University of Chicago Press, 2007), pp. 135–73

Elly Dekker, *Illustrating the Phaenomena: Celestial Cartography in Antiquity and the Middle Ages* (Oxford: Oxford University Press, 2013)

Elly Dekker and Peter van der Krogt, *Globes from the Western World* (London: Zwemmer, 1993)

Elly Dekker and Kristen Lippincott, 'The Scientific Instruments in Holbein's *Ambassadors*: A Re-examination', *Journal of the Warburg and Courtauld Institutes*, 62 (1999), 93–125

Wolfram Dolz, 'Determination of Wood Types in Globe Stands', *Der Globusfreund*, 38/39 (1990/91), 121–30

Johannes Dörflinger, 'Printed Austrian Globes (18th to early 20th centuries)', *Der Globusfreund*, 35/37 (1987), 191–205

Chet Van Duzer, *Johann Schöner's Globe of 1515: Transcription and Study* (Philadelphia: American Philosophical Society, 2010)

Stephen Edell, 'Concave Hemispheres of the Starry Orb', *Bulletin of the Scientific Instrument Society*, 7 (1985), 6–8

Felipe Fernandez-Armesto (ed.), *The Times Atlas of World Exploration* (London: HarperCollins, 1991)

Matteo Fiorini, *Sfere Terrestri e Celesti di Autore Italiano Oppure Fatte o Conservate in Italia* (Rome: Presso la Società Geografica Italiana, 1899)

Sir Herbert George Fordham, *John Cary: Engraver, Map, Chart and Print-Seller and Globe-Maker* (Cambridge: Cambridge University Press, 1925)

Derek Gillard, *Education in England: A Brief History* (www.educationengland.org.uk/history)

Derek Howse, *Greenwich Time and the Discovery of the Longitude* (Oxford: Oxford University Press, 1980)

Nick Kanas, *Star Maps: History, Artistry and Cartography* (Chichester: Praxis, 2009)

Ib Rønne Kejlbo, *Rare Globes* (Copenhagen: Munksgaard/Rosinante, 1995)

Peter Kemp (ed.), *The Oxford Companion to Ships and the Sea* (Oxford: Oxford University Press, 1990)

Cornelis Koeman, Günther Schilder, Marco van Egmond and Peter van der Krogt, 'Commercial Cartography and Map Production in the Low Countries, 1500–ca. 1672', in David Woodward (ed.), *The History of Cartography, Volume Three: Cartography in the European Renaissance Part 2* (Chicago: University of Chicago Press, 2007), pp. 1296–1383

Peter van der Krogt, 'Globes, Made Portable for the Pocket', *Bulletin of the Scientific Instrument Society*, 7 (1985), 8–15

Peter van der Krogt, *Globi Neerlandici: The Production of Globes in the Low Countries* (Utrecht: Hes, 1993)

Peter van der Krogt, *Old Globes in the Netherlands* (Utrecht: Hes, 1984)

Tom Lamb and Jeremy Collins (eds), *The World in your Hands* (Leiden: Museum Boerhaave/London: Christie's, 1994)

Rene Lehmann, *Berliner Globenhersteller 1790–1970* (Berlin: Lehmanns Colonialwaren im Eigenverlag, 2010)

Kristen Lippincott, 'A Chapter in the Nachleben of the Farnese Atlas: Martin Folkes's Globe', *Journal of the Warburg and Courtauld Institutes*, 74 (2011), 281–99

Kristen Lippincott, *The Story of Time* (London: Merrell Holberton/National Maritime Museum, 1999)

Raymond Lister, *Old Maps and Globes* (London: Bell & Hyman, 1979)

Marica Milanesi and Rudolf Schmidt (eds), *Sfere del Cielo Sfere della Terra* (Milan: Electa, 2007)

John R. Millburn, *Adams of Fleet Street, Instrument Makers to King George III* (London: Ashgate, 2000)

John R. Millburn, *Benjamin Martin: Author, Instrument-Maker, and Country Showman* (Leyden: Noordhoff, 1976)

John R. Millburn, *Wheelwright of the Heavens* (London: Vade-Mecum, 1988)

John R. Millburn and Tor E. Rössaak, 'The Bardin Family, Globe-Makers in London, and Their Associate, Gabriel Wright', *Der Globusfreund*, 40/41 (1992/3), 21–58

Jan Mokre, *Rund um den Globus* (Vienna: Bibliophile, 2008)

Oswald Muris and Gert Saarmann, *Der Globus im Wandel der Zeiten* (Berlin: Columbus Verlag Paul Oestergaard, 1961)

Joseph Needham, 'Celestial Globes', in *Science and Civilisation in China*, vol. 3, *Mathematics and the Sciences of the Heavens and Earth* (Cambridge: Cambridge University Press, 1959), pp. 382–90

Joseph Needham, 'Terrestrial Globes', in *Science and Civilisation in China*, vol. 4, part 3, *Civil Engineering and Nautics* (Cambridge: Cambridge University Press, 1971) pp. 584–7

Mary Sponberg Pedley, *Bel et Utile: The Work of the Robert de Vaugondy Family of Mapmakers* (Tring: Map Collector Publications, 1992)

Walter Ristow, 'Maps by Pendleton's Lithography', *The Map Collector*, 21 December 1982, 26–31

Emily Savage-Smith, *Islamicate Celestial Globes: Their History, Construction, and Use* (Washington, DC: Smithsonian Institution Press, 1985)

Rodney W. Shirley, *The Mapping of the World: Early Printed World Maps 1472–1700* (Riverside, CT: Early World Press, 2001)

A. D. C. Simpson, 'Globe Production in Scotland in the Period 1770–1830', *Der Globusfreund*, 35/37 (1987), 21–32

Edward Luther Stevenson, *Terrestrial and Celestial Globes* (New Haven: Yale University Press, 1921)

Lucy Trench (ed.), *Materials and Techniques in the Decorative Arts: An Illustrated Dictionary* (London: John Murray, 2000)

R. V. Tooley, *Maps and Map-Makers* (New York: Bonanza Books, 1961)

Sarah Tyacke, *London Map-Sellers 1660–1720* (Tring: Map Collector Publications, 1978)

Vladimiro Valerio, 'Giovanni Maria Cassini's Globe Gores (1790/1792) – A Study of Text and Images', *Globe Studies*, 51/52 (2005), 73–84

Helen Wallis, 'The First English Globe: A Recent Discovery', *Geographical Journal*, 117 (3), September 1951, 275–90

Helen Wallis, 'Geography is Better than Divinitie', in Norman Thrower (ed.), *The Compleat Plattmaker* (Berkeley: University of California Press, 1978), pp. 1–43

Helen Wallis, 'The Place of Globes in English Education 1600–1800', *Der Globusfreund*, 25/27 (1978), 103–10

Helen Wallis (ed.), *Vincenzo Coronelli, Libro dei Globi 1693 (1701)* (Amsterdam: Theatrum Orbis Terrarum, 1969)

Helen Wallis and E. D. Grinstead, 'A Chinese Terrestrial Globe, A.D. 1623', *British Museum Quarterly*, 25 (314), June 1962, 83–91

Deborah J. Warner, *The Sky Explored: Celestial Cartography 1500–1800* (New York/Amsterdam: Theatrum Orbis Terrarum, 1979)

Diederick Wildeman, *De Wereld in Het Klein: Globes in Nederland* (Amsterdam: Walberg Pers/Vereeniging Nederlandsch Historisch Scheepvaart Museum, 2006)

Johannes Willers, 'Die Geschichte des Behaim-Globus', in *Focus Behaim Globus I* (Nuremberg: Germanisches Nationalmuseum, 1992), pp. 209–16

Laurence Worms and Ashley Baynton-Williams, *British Map Engravers* (London: Rare Book Society, 2011)

Ena L. Yonge, *A Catalogue of Early Globes* (New York: American Geographical Society, 1968)

# PICTURE CREDITS

# INDEX

Figures in *italic* refer to pages on which illustrations appear.